澜湄职业教育培训中心暨柬埔寨鲁班工坊系列教材
A Series of Textbooks for Lancang-Mekong Vocational Education Training Center and Cambodia Luban Workshop

电工——高级

Electrotechnology—Advanced Level

主编：李云龙　张佳　张新江
Chief editor：LI Yunlong　ZHANG Jia　ZHANG Xinjiang

西安电子科技大学出版社

Brief Introduction

This book is based on the "National Vocational Skills Standards—Maintenance Electrician" and is compiled around the concept of "facing the actual conditions of Cambodia, taking the needs of universities, society and enterprises as the orientation, and taking vocational competence as the core". Each module in this book contains theoretical knowledge and operation procedures. The theoretical part introduces the general ideas of theoretical serial lecture, starting from the assessment points of the National Vocational Skills Assessment, being supplemented by detailed analysis of typical examples, aiming to make readers grasp the key points. The operation part uses advanced training equipment as the operating platform to exercise practical skills and demonstrates the practical process of typical cases, which can improve the operation skills. The book is striving to highlight the characteristics of vocational training and meet the needs of vocational training and assessment.

This book can be used not only as a training material for related majors in Cambodian colleges and universities, but also as a learning material for skilled workers who are willing to participate in the advanced assessment of National Vocational Skills Assessment. It can also be used as reference books in on-the-job training.

图书在版编目(CIP)数据

电工=Electrotechnology—Advanced Level：高级：英文 / 李云龙，张佳，张新江主编. — 西安：西安电子科技大学出版社，2023.5
ISBN 978-7-5606-6705-8

Ⅰ. ①电… Ⅱ. ①李… ②张… ③张… Ⅲ. ①电工技术－教材－英文 Ⅳ. ①TM

中国版本图书馆 CIP 数据核字(2022)第 200524 号

策划编辑　明政珠
责任编辑　明政珠
出版发行　西安电子科技大学出版社(西安市太白南路 2 号)
电　　话　(029)88202421　88201467　邮　　编　710071
网　　址　www.xduph.com　电子邮箱　xdupfxb001@163.com
经　　销　新华书店
印刷单位　咸阳华盛印务有限责任公司
版　　次　2023 年 5 月第 1 版　2023 年 5 月第 1 次印刷
开　　本　787 毫米×1092 毫米　1/16　印　张　13
字　　数　301 千字
印　　数　1～1000 册
定　　价　42.00 元
ISBN　978-7-5606-6705-8/ TM
XDUP 7007001-1
***** 如有印装问题可调换 *****

General Foreword

Serving the Belt and Road Initiative of China, the Lancang-Mekong Vocational Education Training Center and Cambodia Luban Workshop is a joint project undertaken by Tianjin Sino-German University of Applied Sciences(TSGUAS) for the Ministry of Foreign Affairs, the Ministry of Education and the Tianjin Municipal People's Government. Based in Cambodia, the project is designed to serve five countries in the Lancang-Mekong area and radiates to other ten ASEAN countries. It integrates functions of vocational training, vocational education, scientific research, cultural inheritance and innovation & entrepreneurship, develops both academic and non-academic education, and operates as a market-oriented international vocational training center.

At the initial stage of the project, 18 training rooms including mechanical processing technology, electrical technology and communication technology were built in three training centers for mechatronics and communication technology related majors, with a total construction area of 6814 m^2 and more than 1600 sets of equipment.

The project will implement a "three-phase" plan. Specialty construction is in the first phase. International tourism, logistics engineering, automobile maintenance, building electricity and other specialties will be set up in the second phase to carry out technical skills training for Chinese & Cambodian enterprises and Cambodian people. Meanwhile, higher vocational education, applied technology undergraduate education, joint postgraduate education and other academic educations will be carried out to explore systematic talents cultivation of "medium and high vocational education, undergraduate education, and postgraduate education for a master's and doctoral degree".

Since 2017, as many as 95 articles about the project have been published by mainstream media including People's Daily, Guangming Daily, China Education News, Xinhuanet, etc. from home and abroad. After over two months of field study and research, Tianjin Television produced two feature stories named "Khmer Training", each lasting 30 minutes. The two episodes were broadcast on May 6th , 2019 and May 13th, 2019 respectively, featuring "on and on sails the vocational education overseas to shine on the Luban Workshop". They give a full coverage of how TSGUAS teachers brought advanced skills to local areas and how friendship flourished along the Belt and Road Initiative—a great contribution to the BRI. On July 18th, 2019, the Royal Government of Cambodia conferred the Officer of the SAHAMETREI Medal to the Secretary of the Party Committee of TSGUAS, and the Knight of the SAHAMETREI Medal to the President and Vice President in charge of this project, with the signature of Prime Minister Hun Sen of Cambodia. On July 22, 2019, China Education Association for International Exchange awarded TSGUAS the medal of "Featured Cooperation Project of China-ASEAN Higher Vocational Colleges". In October, 2019, the President of National Polytechnic Institute of Cambodia (NPIC) presented 11 teachers with certificates and medals for their outstanding contributions to the Ministry of Labor and Vocational Training of Cambodia. Tianjin Sino-German University of Applied Sciences together with National Polytechnic Institute of Cambodia (NPIC) and their partners with enterprises was approved as the Belt and Road Joint Laboratory (Research Center)—Tianjin Sino-German and Cambodia Intelligent Motion Device and Communication Technology Promotion Center in December, 2020.

The Center has become a training base in Lancang-Mekong areas for technical talents training, a talent support base for Chinese enterprises overseas, a demonstration base for international students, and a base for teachers training. The Center is a key educational project of the Ministry of Foreign Affairs to serve the Belt and Road Initiative with foreign participation and entity institutions involved locally. The project will serve the social-economic and cultural development of the countries along with the Initiative, enhancing the well-being of mankind; it will also serve the production output capacity of Chinese enterprises to help national development as well as enhance the international development of vocational education and the quality of its connotation. The project is a bridge connecting vocational education of Tianjin with the world, which marks a new stage of the city's international exchange and cooperation from a lower-medium to a medium-higher level.

The team of the project has compiled a series of textbooks for training, involving six occupations (electrotechnology, lathe, milling, CNC operation, bench and 4G communication network) from elementary, intermediate to advanced level based on current human resources situation in Lancang-Mekong countries, China's teaching equipment, and Chinese vocational qualification standards. These 19 textbooks target competence development and orient students to work tasks, combining theory with practice, and learning with practicing so as to put knowledge and skills into real situations. The textbooks aim to provide skills standards for the six occupations and lay foundations for the upgrading of the technological level of Lancang-Mekong countries.

ZHANG Xinghui

Party Secretary of Tianjin Sino-German University of Applied Sciences

March, 2021

A Series of Textbooks for Lancang-Mekong Vocational Education Training Center and Cambodia Luban Workshop Editorial Committee

Preface

In order to better serve China's "The Belt and the Road Initiative" and Lancang-Mekong Vocational Education and Training Center, and export high-quality technical skills training and educational concepts to Lancang-Mekong countries, the teachers of the Lancang-Mekong Vocational Education and Training Center project team compiled this book based on China's vocational skill standards and advanced training equipment. The characteristic of this book is as follows. This book is based on the technical skills and theoretical knowledge requirements of the advanced level of the national vocational qualification standards for "electricians" and on the most advanced training equipment in the current domestic vocational education field, and it flexibly uses the scientific education theory of the work process system to integrate what knowledge and skills students need in each module and specific task. This book has made scientific and reasonable innovations and breakthroughs in content and structure. It is compiled focusing on ability and actual tasks, and according to the principles of project-based teaching mode, which combines theory with practice, makes students learn while practicing according to actual task requirements, and truly apply knowledge and skills.

The book has designed five modules including Maintenance and troubleshooting of common electrical control circuits, Comprehensive training of power electronic circuit technology, Comprehensive training of electrical drive system, Comprehensive training of programmable logic control system, and Programming and debugging of YL-335B automatic production line. There are two to four specific work tasks under each module, and each task starts with the basic theoretical knowledge required to complete the task, which combines theory and practical work cases. At the same time, this book provides a number of extended practice for readers who have extra ability to learn.

The modules and work content designed in this book are derived from actual production cases and can meet all the requirements and regulations of the national professional standards for "electrician" on elementary theoretical knowledge and related skills. This book can be used as a teaching book for Electrical Automation, Mechatronics, Energy Engineering, Building Electrical and other related majors, or as reference books or skills training books for technical personnel. At the same time, the book is written in English with Chinese version in QR code, and is in line with bilingual education materials. It can not only meet the needs of domestic readers for electrical related learning, but also can be applied to overseas related skills training and foreign readers' self-study.

The book is chief-edited by LI Yunlong, ZHANG Jia and ZHANG Xinjiang of Tianjin Sino-German University of Applied Sciences, and co-edited by ZHANG Lian, NIU Yunsheng, and YANG Zhongli of Tianjin Sino-German University of Applied Sciences. Its English version is worked by ZHENG Yilin of Tianjin Sino-German University of Applied Sciences. LI Yunlong compiled Modules 3, 4 and 5; ZHANG Xinjiang compiled Module 1, and ZHANG Jia compiled Module 2 and 3. ZHANG Lian compiled the Appendix 1 and Appendix 2. Niu Yunsheng and Yang Zhongli (Tianjin Sino-German University of Applied Sciences) gave guidance on the style and content, and provided some references for this book.

Due to the limited abilities of the editors, it is welcomed to provide your suggestions and advices.

Editors

Feb. 20th, 2020

Contents

Module 1 Maintenance and troubleshooting of common electrical control circuits

【General Introduction】

In the industrial sector, the knowledge of the maintenance and troubleshooting of typical electrical control circuits is widely used and it is an important technical means for production and processing. By studying the maintenance and troubleshooting of the electrical control circuit of the CA6140 lathe, the electrical control circuit of the electric hoist, the electrical control circuit of the M7120 surface grinder, and the maintenance and troubleshooting of Z3050 radial drilling machine circuit, students can carry out the debugging, maintenance and inspection of the typical electrical control system, and apply electrical control circuit troubleshooting and inspection skills to other industrial equipment through summarizing.

【Training Objectives】

1. Knowledge objectives

(1) Understand the model, structure, basic function and purpose of commonly-used electrical control equipment.

(2) Master the installation methods and steps of commonly-used electrical control circuits.

(3) Master the debugging methods and steps of commonly-used electrical control circuits.

(4) Master the safety operating procedures of commonly-used electrical control equipment.

2. Competence objectives

(1) Cultivate the ability of self-determined learning, independent analysis and problem-solving.

(2) Cultivate strong hands-on skills.

(3) Cultivate strong team organization and collaboration capabilities.

(4) Increase awareness of standards and regulations, safe production, quality, environmental protection, and service.

(5) Cultivate the ability of knowledge expansion and innovation.

3. Operating standards

(1) Able to wire and select the commonly-used electrical control equipment motors and electrical components.

(2) Able to simply analyze the principles of commonly-used electrical control equipment systems.

(3) Able to draw schematic diagrams and wiring diagrams of commonly-used electrical control equipment circuits.

(4) Able to read the electrical schematic diagrams and electrical component layout diagrams of commonly used electrical control equipment circuits.

(5) Able to select models of relevant low-voltage electrical components.

(6) Able to energize and debug according to the process requirements.

(7) Able to comply with safe operation regulations.

4. Safety standards

In order to ensure the personal safety and equipment safety during the training, the participants must strictly abide by the following safety operating procedures.

(1) The participants should ensure the training devices having reliable protective grounding, leakage protection, overload protection, insulated table tops, insulated carpets and safety patch cords.

(2) During training, the human body should not be in contact with bare live objects.

(3) When in wiring and disconnecting, the equipment must with the power supply cut off.

(4) When for switching in, all participants of the training team must acknowledge and agree.

(5) When encountering abnormal situations, for example, peculiar smell (such as the burning smell of varnish), the electrical unit overheated or vibrating (or too much noise during running), the fuse burned out, sparks appeared in the unit, and the meter over full for measuring range (or reverse bias), the experimenter should cut off the power immediately, find the cause and eliminate the fault.

(6) Before energizing the unit, check whether the motor can rotate flexibly. (Check if the rotor is stuck). If it cannot rotate flexibly, it should be re-adjusted so that the rotor can rotate flexibly.

(7) The total power supply should be controlled by the training instructor. Other personnel can only operate with the permission of the instructor. They are not allowed to switch in at their option.

(8) When using the rheostat, it cannot be set to zero (or a position with a small resistance value), as it is likely to burn the varistor. In the experiment, adjust the high-value resistance first, and then adjust the low-value resistance.

5. Quality standards

Training instructor should evaluate the students' attitude, work quality, operating specifications in the work process by oral examination. The focus of the assessment is to evaluate whether the students have the professional ability of repairing electricians, whether they can meet the professional job requirements of the installation, debugging, maintenance and troubleshooting of electrical system circuits and devices, and whether they have good occupational qualities such as proactivity and enthusiasm.

【Training Requirements】

1. Preparations of tools, instruments and equipment

1) Power supply

(1) Single-phase AC and DC adjustable power supply. After the three-phase AC power passes through the three-phase isolation transformer, a set of 220V AC voltage is the output. Through the single-phase voltage regulator, it becomes a single-phase adjustable power supply with an adjustment range of AC 0–220V. The power supply is equipped with short-circuit protection and overload protection. In addition, this single-phase adjustable AC power is converted into a DC adjustable power supply with an adjustment range of DC 0–220V after bridge rectification and capacitor filtering. It has short-circuit protection and displays the output voltage with a digital voltmeter.

(2) Various AC power sources. Transformers can output alternating current of different voltages. The primary side of the transformer can be added with 220V or 380V AC power according to different wiring. Turn on the switch, and the secondary side of the transformer can output 110V, 36V, 20V, 24V, 12V, 6.3V AC voltage.

(3) DC adjustable regulated power supply. The adjustable regulated power supply is composed of a DC voltage stabilizing circuit and potentiometer, and the potentiometer can be adjusted to obtain a DC power supply.

(4) DC stabilized power supply. Turn on the switch of DC stabilized power supply, and the stable DC power can be obtained.

2) Instrument

1 DC digital voltmeter, 1 DC digital ammeter, 1 AC digital voltmeter, 1 AC digital ammeter, 1 AC power meter; The basic configurations of other electrical components and tools are shown as in Table 1-1.

Table 1-1 Basic configurations of other electrical components and tools

No.	Name	Unit	Amount
1	Three-phase leakage switch	Piece	1
2	Fuse	Piece	2
3	3P fuse	Piece	2
4	Master switch	Piece	7
5	AC contactor	Piece	3
6	Thermal relay	Piece	2
7	Three-phase AC asynchronous motor	Set	3
8	Terminal board, wire slot, wire	—	Appropriate amount for each
9	Screwdriver	Piece	1
10	Electrician's pliers	Piece	1
11	Wire stripper	Piece	1
12	Needle nose pliers	Piece	1
13	Multimeter	Piece	1

2. Knowledge assessment points

(1) The working principle of the electrical control circuit of CA6140 lathe.

(2) The working principle of the electrical control circuit of electric hoist.

(3) The working principle of the electrical control circuit of M7120 surface grinder.

3. Skill assessment points

(1) Analysis and maintenance of fault phenomena of electrical control circuit of CA6140 lathe.

(2) Analysis and maintenance of fault phenomena of electric control circuit of electric hoist.

(3) Analysis and maintenance of fault phenomena of electrical control circuit of M7120 surface grinder.

【Work Tasks】

Task 1 Maintenance and troubleshooting of common faults in the electrical control circuit of CA6140 lathe

✍ Relevant Knowledge

CA6140 is a horizontal lathe improved on the basis of the original C620. C stands for lathe and A is the structural feature code to distinguish CA6140; 6 stands for horizontal; 1 stands for basic type, and 40 stands for maximum rotation diameter.

1. The composition and function of CA6140 lathe

(1) Spindle box. It is fixed on the left end of the machine body. The main shaft in the spindle box is hollow being used for the processing of longer bars and the laying of machine tool lines, such molding can increase the rigidity of the main shaft. The workpiece is clamped by clamping chuck. The function of the spindle box is to support and drive the main shaft, so that the main shaft drives the workpiece to rotate at the specified speed.

(2) Bed saddle and tool rest components. The bed saddle is located in the middle of the bed and can move longitudinally along the tool post track on the bed. The tool rest part is located on the saddle, and its function is to clamp the turning tool and make the turning tool move longitudinally, horizontally or obliquely.

(3) Tailstock. It is located on the tailstock rail of the bed, and its position can be adjusted longitudinally along the rail. The function of the tailstock is to support the workpiece with the rear center. Drills and other processing tools can also be installed on the tailstock for hole processing.

(4) Feed box. It is fixed on the left front side of the bed and the bottom of the spindle box. Its function is to change the pitch of the machined thread or the amount of motorized feed.

(5) Slide box. It is fixed at the bottom of the tool holder component, and can drive the tool holder to do longitudinal, horizontal feed, rapid movement or thread processing together. Various

operating handles and buttons are installed on the slide box, so that workers can conveniently operate the machine tool during work.

(6) Bed. The bed is fixed on the left leg and the right leg. The bed is the basic support of the machine tool. The main parts of the machine tool are installed on the bed, and the bed keeps them in an accurate relative position when working.

2. The main drive of the machine tool

The function of the main drive of the machine tool is to transmit the speed and torque of the motor to the spindle through a certain way, so that the workpiece can move at different speeds. The performance of the main drive directly affects the processing quality and production efficiency of the parts. Taking into account the economy of transformation, the main drive of the machine is an original ordinary three-phase asynchronous AC motor. Considering that when the grid voltage and cutting torque change during the machining process, the speed of the motor will also fluctuate, which directly affects the surface roughness of the machined parts. Therefore, in order to improve the machining accuracy and realize the automatic stepless speed change of the main shaft, an asynchronous AC motor frequency conversion speed regulation system is added to the main shaft, thereby eliminating the need for mechanical gear shifting.

3. Analysis of main circuit

There are three motors in the main circuit; M1 is the spindle motor, which drives the rotation of the spindle and the tool post for feed movement; M2 is the cooling pump motor; M3 is the fast moving motor of the tool post.

The three-phase AC power supply is introduced through the transfer switch QS1. The spindle motor M1 is controlled and started by the contactor KM1, and the thermal relay FR1 is the overload protection of the spindle motor M1. The cooling pump motor M2 is controlled and started by the contactor KM2, and the thermal relay FR2 is its overload protection. The rapid movement motor M3 of the tool post is controlled and started by the contactor KM3.

4. Analysis of control circuit

The power of the control loop is provided by the 110V voltage output from the secondary side of the control transformer TC.

(1) Control of the spindle motor. Press the start button SB1, and the coil of the contactor KM1 is energized. Its main contact is closed, and the spindle motor starts to run. At the same time, the self-locking contact of KM1 and another pair of normally open contacts are closed. Press the stop button SB2 to stop the spindle motor M1.

(2) Cooling pump motor control. If the process needs to use coolant during the turning process, close the switch SA1. When the spindle motor M1 is running, the contactor KM1 coil is energized and the main contact is closed, and the cooling pump motor M2 is energized. It can be seen from the electrical schematic diagram that the cooling pump motor M2 is only possible to start after the spindle motor M1 starts. When M1 stops running, M2 also stops automatically.

(3) The control of the fast moving motor of the tool post. The start of the rapid-moving

motor M3 of the tool post is controlled by the button SB3 installed at the top of the feed joystick, such control forms the jog control link with the intermediate relay KM2. Move the joystick to the desired direction, press the button SB3, the relay KM2 is energized, and the M3 is activated, and the tool post moves quickly in the specified direction.

5. Analysis of lighting and signal lamp circuits

The secondary side of the control transformer TC outputs voltages of 24 V and 6 V respectively, which are used as the power supply for the low-voltage lighting lamps and signal lamps of the machine tool. EL is the low-voltage lighting lamp of the machine tool, which is controlled by the switch SA; HL is the signal lamp of the power supply. They use FU4 and FU3 for short-circuit protection respectively. Figure 1-1 is the schematic diagram of CA6140 lathe.

✍ Practice and Operation

1. Task preparations

Before the task, prepare the basic tools and accessories used in the training, as shown in Table 1-2.

Table 1-2　Common troubleshooting and inspection tools and accessories for the electrical control circuit of the CA6140 lathe

No.	Name	Model specification	Unit	Amount
1	Three-phase leakage switch	DZ47-60 10A	Piece	1
2	Fuse	RL1-15	Piece	2
3	3P Fuse	RT18-32	Piece	2
4	Master switch	LS1-1	Piece	7
5	AC contactor	CJ20-10	Piece	3
6	Thermal relay	JR36-20	Piece	2
7	Three-phase AC asynchronous motor		Set	3
8	Terminal board, wire slot, wire			Appropriate amount for each
9	screwdriver		Piece	1
10	Electrician's pliers		Piece	1
11	Wire stripper		Piece	1
12	Needle nose pliers		Piece	1
13	Multimeter		Piece	1

2. Task content

Proceed with the maintenance and troubleshooting of common faults in the electrical control circuit of the CA6140 lathe. The schematic diagram of CA6140 lathe is as Figure 1-1.

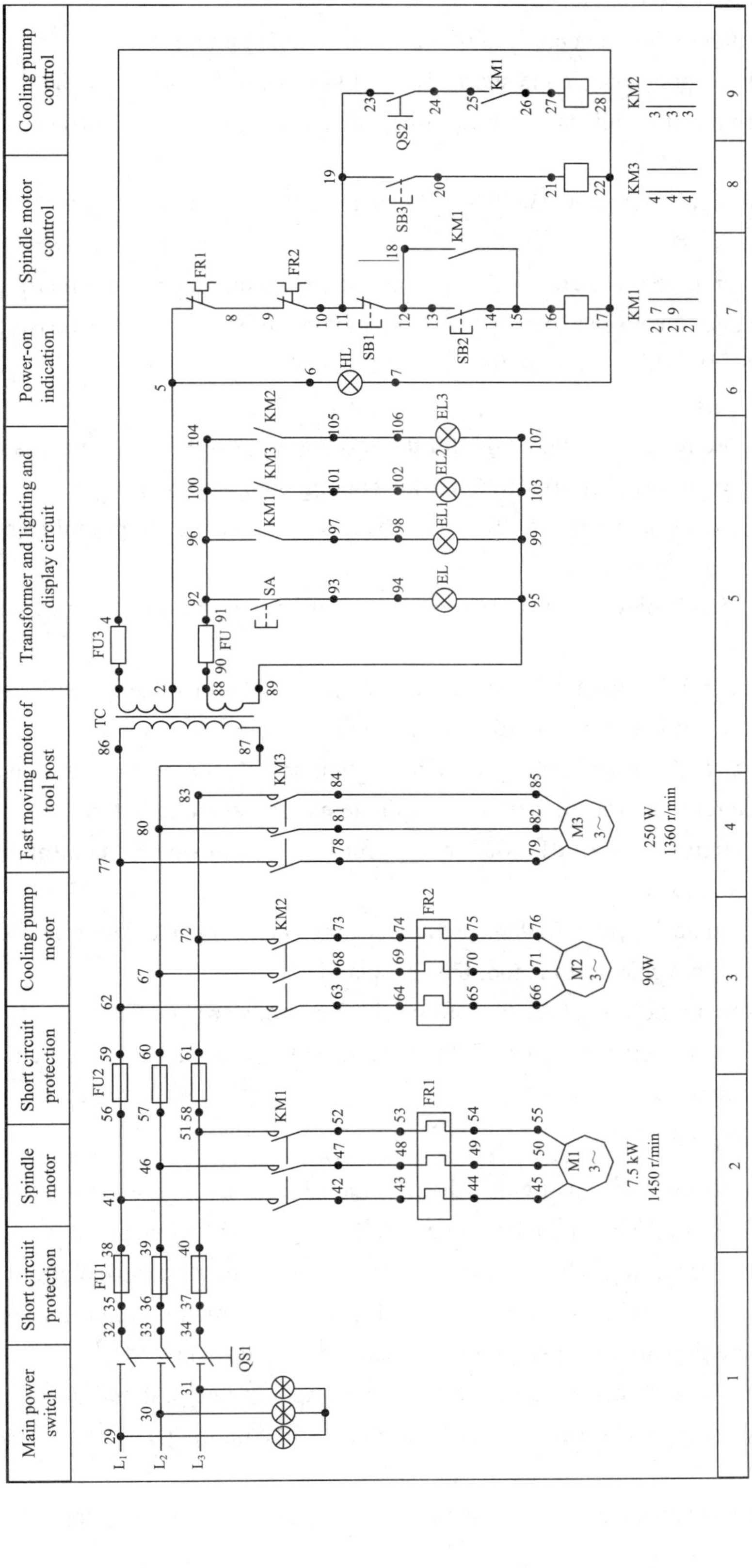

Figure 1-1 Schematic diagram of CA6140 lathe

3. Task procedures

(1) Record the failure phenomena of the circuit training unit of the CA6140 lathe.

(2) Check the open circuit and short circuit of the machine tool circuit and make a record.

(3) Determine the fault location according to the following fault phenomenon comparison details:

① Open circuit between 038–041: for all motors, one phase is missing, and all control circuits fail.

② Open circuit between 049–050: for the spindle motor, one phase is missing.

③ Open circuit between 052–053: for the spindle motor, one phase is missing.

④ Open circuit between 060–067: for M2 and M3 motors, one phase is missing, and the control circuit fails.

⑤ Open circuit between 063–064: for the cooling pump motor, one phase is missing.

⑥ Open circuit between 075–076: for the cooling pump motor, one phase is missing.

⑦ Open circuit between 078–079: one phase of the motor for rapid movement of the tool post is missing.

⑧ Open circuit between 084–085: one phase is missing in the fast moving motor of the tool post.

⑨ Open circuit between 002–005: except the lighting lamp, other controls are invalid.

⑩ Open circuit between 004–028: control circuit failure.

⑪ Open circuit between 008–009: indicator light is on, other controls are invalid.

⑫ Open circuit between 015–016: the spindle motor cannot be started.

⑬ Open circuit between 017–022: except for the rapid movement control of the tool post, other controls are invalid.

⑭ Open circuit between 020–021: the fast movement motor of the tool post is unable to start, and the fast movement of the tool post is invalid.

⑮ Open circuit between 022–028: all machine controls are invalid.

⑯ Open circuit between 026–027: the spindle motor starts; the cooling pump control fails, and QS2 does not work.

4. Precautions

(1) Establish groups and divide labour reasonably. Trainings are usually carried out in groups, each consist of 2 to 3 people. The training usually involves wiring, load adjusting, meter reading, and data recording. The members of the training team must have a clear and reasonable division of labor, coordinate their work, and conduct continuous consultations during the training to ensure the training operation is standardized and coordinated.

(2) Select unit and meter. Before wiring, select appropriate unit and instrument (including the range of the instrument) according to the purpose of the training, and record their models and parameters.

(3) Numerical value reading and abnormality handling. When reading the meter, pay attention

to the magnification, minimum scale value, unit and significant digits. If an abnormality is found, immediately cut off the power supply, find the cause, and eliminate the fault.

After the training is completed, conduct a preliminary analysis of the training data to check whether there are any abnormalities. If there are any, the relevant data at this point should be re-examined. If any errors are found, they should be corrected.

5. Notes and summary

____

____

____

____

____

____

Task 2 Maintenance and troubleshooting of common faults in the electric control circuit of electric hoist

✍ Relevant Knowledge

1. Overview of electric hoist

Electric hoist is a kind of special lifting equipment, installed on overhead cranes and gantry cranes. The electric hoist has the characteristics of small size, light weight, and simple operation procedure. It is used in industrial and mining enterprises, storehouse, docks and other places.

The electric hoist has a compact structure. The hydraulic system is dual-controlled, and the pressure can be precisely controlled by the overflow regulating valve and the magnetic contact pressure gauge. The electrical control part adopts low-voltage control, which increases the safety of the control system.

The lifting weight of the electric hoist is generally 0.3 to 80 tons, and the lifting height is 3 to 30 meters. Being composed of a motor, a transmission mechanism and a drum or sprocket, the electric hoist is divided into two types: electric chain hoist and wire rope electric hoist. Among them, electric chain hoists are divided into two types: imported and domestic; wire rope electric hoists are divided into single-speed hoist and double-speed hoist. Other electric hoists also include miniature electric hoist, windlass, and multi-function hoist.

For one electric hoist, parts such as reducer, hoisting motor, running motor, flame interrupter, cable slip line, reel device, hook device, coupling, flexible cable current introducer and etc. integrate power and braking force in each of them.

According to the relative position between the motor and the drum for each kind electric hoist, it can be roughly divided into following types:

(1) The electric hoist with its motor axis perpendicular to the drum axis. It adopts worm gear transmission with characteristics of large width, cumbersome structure, low mechanical efficiency. Because its processing is difficult, this type of electric hoist went out of production now.

(2) The electric hoist with the motor axis parallel to the drum axis. It has the advantage of small height and length. Its disadvantages are large width, poor grouping, complex manufacturing and assembly, and large track turning radius.

(3) The electric hoist with the motor installed in the drum. It has the advantages of small length and compact structure. The disadvantages of it are poor heat dissipation conditions of its motor, poor grouping, inconvenient in inspection, installation and maintenance of its motor, and with complicated power supply devices.

(4) The electric hoist with the motor mounted on the outside of the drum. It has the advantages of good grouping, high degree of generalization, easy to change the lifting height, and convenience in installation and maintenance. The disadvantage is that the length of the electric hoist is large.

The structure and characteristics of the wire rope electric hoist: the machine adopts an electromechanical integration design. By changing different molds, you can compact different specifications of wire ropes. It is easy and safe to control and convenient in inspection, installation and maintenance of its motor. The main disadvantages are the poor heat dissipation of its motor, poor grouping, and complicated power supply devices, complex in manufacturing and assembly.

The components of wire rope electric hoist are motor, transmission gear, drum and sprocket. The steel wire rope achieved through compacting has large tensile force and fast speed. It only takes a few minutes to crimp a set of steel wire rope, which greatly reduces labor intensity and improves work efficiency.

Electric hoists are widely used, but there are fewer professionals who can correctly operate electric hoist. It is recommended that you learn more about electric hoist and make better use of various electric hoist in various fields.

2. Electric control circuit of electric hoist

The schematic diagram of electric control circuit of electric hoist is shown in Figure 1-2. In the main circuit, when the main contact point of the contactor KM1 is closed, the motor stator winding is in forward (up); when the main contact point of the contactor KM2 is closed, the motor stator winding is in reverse (down). The thermal element of the thermal relay FR1 is connected in series with the motor M1 to protect it against overload. When the main contact of KM3 is closed, the motor stator winding is in forward (forward); when the main contact of contactor KM4 is closed, it is in reverse (reverse). The thermal element of the thermal relay FR2 is connected in series with the motor M2 to protect it against overload.

In the auxiliary circuit, pull the switch SQ up, the electromagnetic coil of the contactor KM1 is energized, and the main contact point is closed to cause the motor M1 in forward rotation (rise). When the normally closed auxiliary contact point is disconnected and interlocked, the HL2 indicator light is off; when the normally open auxiliary contact point is closed, HL1 indicator light is on.

After the KM1 normally open auxiliary contact point is closed, pull the switch SQ down, the solenoid coil of the contactor KM2 is energized, the main contact is closed, and the motor M1 is in reverse rotation(drop); the normally closed auxiliary contact opens the interlock, and the HL1 indicator light goes out, the normally open contact is closed, and the HL2 indicator light is on.

To pull the switch SQ left, the solenoid coil of the contactor KM3 is energized, the main contact point is closed, so that the motor M2 is rotating and moving to the left; the normally closed auxiliary contact point is disconnected and interlocked, the HL4 indicator light is off, and the normally open auxiliary contact point is closed, HL3 The indicator light is on.

After the KM2 normally open auxiliary contact is closed, to pull the switch SQ right, the electromagnetic coil of the contactor KM4 is energized, the main contact is closed, the motor M2 reversely rotates and moves to the right; the normally closed auxiliary contact opens the interlock, and the HL3 indicator light is off, the normally open auxiliary contact is closed, and the HL4 indicator light is on.

If close the switch QS, the electromagnetic coils of the contactors KM1, KM2, KM3, and KM4 are de-energized, the main contacts are disconnected, and the two motors stop rotating at the same time; if one or two of the two motors are overloaded, the constant contacted point of the thermal relay FR is disconnected, and both motors stop rotating.

The travel switches SQ1, SQ2, SQ3, SQ4 play a protective role to prevent the motor from exceeding the travel range when it rotates.

✍ Practice and operation

1. Task preparations

Prepare the basic tools and accessories used in the training, as shown in Table 1-3.

Table 1-3 Basic tools and accessories for common troubleshooting of electric hoist electrical control circuit

NO.	Name	Model specification	Unit	Amount
1	Three-phase leakage switch	DZ47LE-32	Piece	1
2	Fuse	RL1-15	Piece	3
3	3P fuse	RT18-32	Piece	1
4	Cross switch	LS1-1	Piece	1
5	Limit switch	LX19-001	Piece	4
6	AC contactor	CJ20-10	Piece	4
7	Thermal relay	JR36-20	Piece	2
8	Three-phase AC asynchronous motor		Set	2
9	Terminal board, wire slot, wire			Appropriate amount for each

Continued table

NO.	Name	Model specification	Unit	Amount
10	Screwdriver		Piece	1
11	Electrician's pliers		Piece	1
12	Wire stripper		Piece	1
13	Needle nose pliers		Piece	1
14	Multimeter		Piece	1

2. Training content

Carry out the maintenance and troubleshooting of common faults in the electric control circuit of the electric hoist.

3. Task procedures

Task procedures are as follows:

(1) Record the failure phenomena of training unit of the electric hoist control circuit.

(2) Detect and record the open circuit and short circuit of the control circuit.

(3) Determine the fault location according to the following fault and symptom comparison details.

① Open circuit between 053-056: for both the hook motor and the moving motor, one phase is missing.

② Open circuit between 057-060: for both the hook motor and the moving motor, one phase is missing, and the display and control circuits are both invalid.

③ Open circuit between 061-081: when the hook is raised, the brake electromagnet is invalid; when the hook is lowered, for the hook motor, one phase is missing.

④ Open circuit between 089-090: for the mobile motor, one phase is missing.

⑤ Open circuit between 093-102: when moving the motor backward, one phase is missing.

⑥ Open circuit between 098-104: display and control circuit failure.

⑦ Open circuit between 005-006: the indicator light is off, and the control circuit is invalid.

⑧ Open circuit between 015-016: press SB1, KM1 coil does not pull in.

⑨ Open circuit between 017-018: press SB1, KM1 coil does not pull in.

⑪ Open circuit between 019-027: the control of hook lowering and forward and backward movement is invalid.

⑫ Open circuit between 025-026: press SB2, KM2 coil does not pull in.

⑬ Open circuit between 027-037: the forward and backward movement control is invalid.

⑭ Open circuit between 033-034: press SB3, KM3 coil does not pull in.

⑮ Open circuit between 035-036: press SB3, KM3 coil does not pull in.

⑯ Open circuit between 037-047: the backward movement control is invalid.

⑰ Open circuit between 045-046: press SB4, KM4 coil does not pull in.

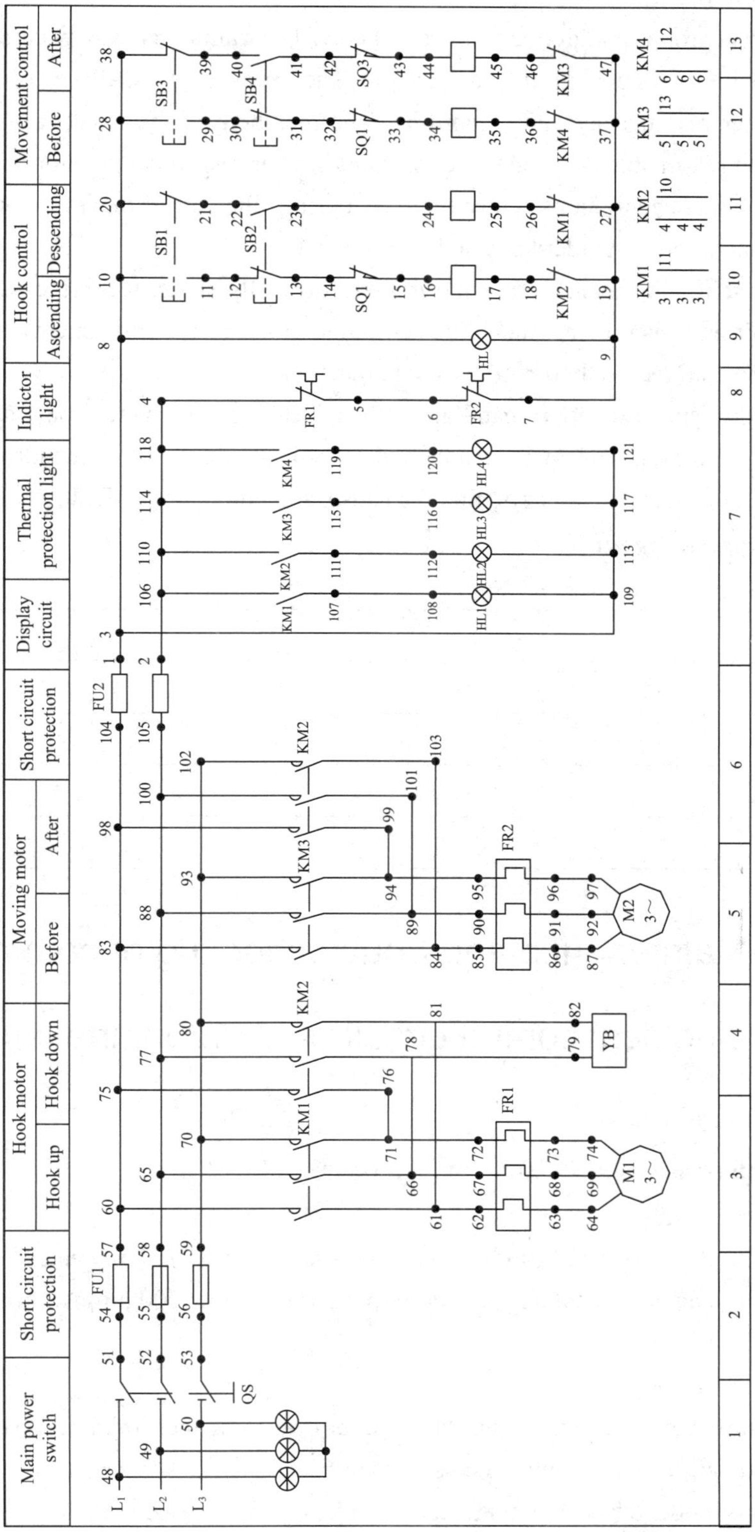

Figure 1-2 Schematic diagram of electric control circuit of electric hoist

4. Precautions

(1) Establish groups and divide tasks reasonably. The training is usually carried out in small groups, and each group consists of 2 to 3 people. The experiment usually involves wiring, load regulation, meter reading, and data recording. The members of the experimental team should have a clear and reasonable division of tasks, stand in their respective positions, coordinate their work, and conduct continuous consultations during the experiment to ensure that the experimental operation is standardized and coordinated.

(2) Select unit and meter. Before the training officially starts, it is necessary to select the appropriate unit and instrument (including the range grade of the instrument) according to the training purpose, and record their models and parameters.

(3) Reading and exception handling. When reading the meter, pay attention to the magnification, minimum scale value, unit and significant digits. If an abnormality is found, immediately cut off the power supply, find the cause, and eliminate the fault.

5. Notes and summary

____

____

____

____

____

____

Task 3 Maintenance and troubleshooting of common faults in the electrical control circuit of M7120 surface grinder

✍ Relevant information

1. Basic overview of M7120 surface grinder

1) Model meaning

For M7120: M represents grinding machine (machine tool); 7 represents plane (working surface); 1 represents horizontal axis counter type; 20 represents 200 mm working surface width of worktable.

2) Basic overview

Grinding machine is a precision machine tool for machining the workpiece around the grinding wheel or the end face. It can not only process ordinary metal materials, but also process high hardness materials such as hardened steel or cemented carbide, and has a wide range of applications.

The M7120 surface grinder has four motors, which are the grinding wheel motor, the grinding wheel lifting motor, the hydraulic pump motor and the cooling pump motor. The grinding wheel motor (providing the power of the grinding wheel) is used to directly drive the grinding wheel to rotate and grind the workpiece. It is generally not required to auto-active the speed of the grinding wheel in the M7120 surface grinder, so the cage asynchronous motor is usually used to drive it. The main movement in the surface grinder is the movement of grinding wheel. The grinding wheel lifting motor makes the grinding wheel move vertically on the column guide rail to adjust the position distance between the grinding wheel and the workpiece. The reciprocating movement of the worktable and the grinding wheel is realized by the hydraulic transmission of the hydraulic pump motor. The hydraulic transmission is relatively stable, which can realize stepless adjustment and stable commutation, and ensure machining accuracy. The cooling pump motor drives the cooling pump to supply the grinding wheel and the grinding workpiece cooling liquid. At the same time, the cooling liquid is used to take away the ground iron filings to prevent grinding pollution and ensure the grinding environment.

2. The main structure and function of M7120 surface grinder

1) The structure of the M7120 surface grinder

The structure of the M7120 surface grinder is shown in Figure 1-3.

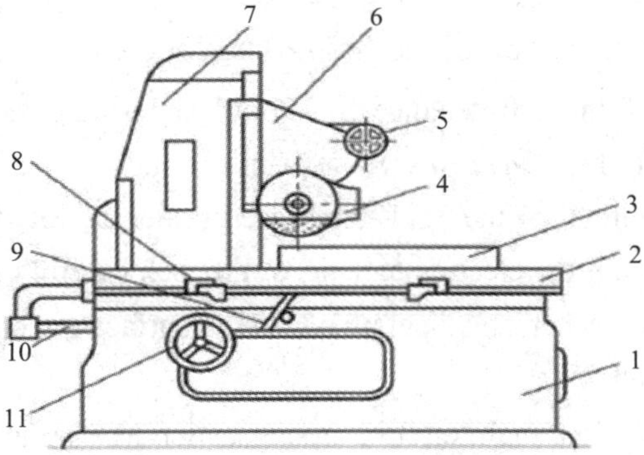

1—Machine bed; 2—Worktable; 3—Electromagnetic chuck; 4—Grinding wheel holder; 5—Laterally moving hand wheel for grinding wheel box; 6—Sliding seat; 7—Column; 8—Worktable reversing striker; 9—Worktable reciprocating reversing handle; 10—Piston rod; 11—Vertical feed handwheel for grinding wheel box

Figure 1-3 Structure of the M7120 surface grinder

2) The components and function of M7120 surface grinder

(1) Grinding wheel holder is used to install the grinding wheel and drive the grinding wheel to rotate at a high speed. The grinding wheel holder can move along the dovetail guide rail of the sliding seat for manual lateral clearance or hydraulic lateral clearance.

(2) Sliding seat is used to install the grinding wheel stand and to drive the grinding wheel stand to move up and down along the column guide rail.

(3) Column is used to support sliding seat and grinding wheel stand.

(4) Workbench is used to install the workpiece and drive it to make reciprocating linear motion by the hydraulic system. The effective working area is 630 mm×200 mm(the working surface width of the workbench). The working surface and the side surface grinding of the

central T-slot are finely processed, with high accuracy and smoothness. They are important base surfaces for installing workpieces or fixtures. They should be used and protected carefully.

(5) Machine bed is used to supports the workbench and installs other structural components for the installation of columns, workbenches, hydraulic systems, electrical components and other operating structures.

(6) Coolant system is used to provide coolant (saponified oil) to the grinding area.

(7) Hydraulic drive system includes the followings.

① Power component includes the oil pump, which supplies pressure oil to the hydraulic transmission system.

② Actuating component includes the oil cylinder, which drives the movement of parts such as the workbench.

③ Control components includes all kinds of valves, which control pressure, speed, direction, etc.

④ Auxiliary components includes fuel tanks, pressure gauges, etc.

Compared with mechanical transmission, hydraulic transmission has the advantages of smooth transmission, overload protection, and stepless speed regulation in a larger range.

3. The electrical control circuit of the M7120 surface grinder

Before proceeding with the maintenance and troubleshooting of common faults, we discuss the electrical control circuit of the M7120 surface grinder.

The electrical control circuit of the M7120 surface grinder can be divided into four parts: the main circuit, the control circuit, the electromagnetic workbench control circuit, and the lighting and indicator circuit. The electrical schematic diagram of M7120 surface grinder is shown in Figure 1-4.

1) Main circuit analysis

There are four motors in the main circuit, among which M1 is the hydraulic pump motor to realize the reciprocating movement of the worktable. M2 is the grinding wheel motor, which drives the grinding wheel to rotate to complete the grinding process. M3 is the cooling pump motor, it only requires one-way rotation, controlled by respectively contactor KM1 and KM2. The cooling pump motor M3 can only run after the grinding wheel motor M2 is running. M4 is the grinding wheel lifting motor, which is used to adjust the position between the grinding wheel and the workpiece during the grinding process.

M1, M2 and M3 usually have long work time, so they are all equipped with overload protection. M4 is for short-term work and does not have overload protection. Four motors share a set of fuse FU1 for short-circuit protection.

2) Control circuit analysis

(1) Control of hydraulic pump motor M1. After closing the main switch QS1, one secondary side of the rectifier transformer outputs 130 V AC voltage, which is rectified by the uni-directional bridge rectifier to obtain a DC voltage, so that the voltage relay KA is energized, and its normally open contact (zone 7) is closed to prepare for the start of the motor. If KA

cannot operate reliably, all motors cannot run. Because the workpieces of the surface grinder rely on the suction force of the DC electromagnetic chuck to firmly hold the workpieces on the worktable, the grinding wheel and hydraulic system are allowed to be started only when a reliable DC voltage is available to ensure safety.

When KA is closed, press the start button SB3. The contactor KM1 is energized and self-locked. The workbench motor M1 starts to run, and the HL2 light is on. If the stop button SB2 is pressed, the contactor KM1 coil is de-energized and released, and the motor M1 is de-energized and stops.

(2) Control of grinding wheel motor M2 and cooling pump motor M3. Press the start button SB5, the contactor KM2 coil is energized, and the grinding wheel motor M2 starts to run. Because the cooling pump motor M3 and M2 are controlled in linked way, M3 and M2 start to operate at the same time. When the stop button SB4 is pressed, the contactor KM3 coil is de-energized and released, and M2 and M3 are de-energized and stopped at the same time.

The normally closed contacts of the thermal breakers FR2 and FR3 of two motors are connected in series in KM2. As long as one motor is overloaded, KM2 will lose power. Because the coolant is circulated, it is often mixed with dirt and impurities, which can easily cause the cooling pump motor M3 to overload, so the thermal relay FR3 is used for overload protection.

(3) Control of grinding wheel lifting motor M4. The grinding wheel lifting motor is only used when its adjusting position between the workpiece and the grinding wheel, so it adopts inching control. When the button SB6 is pressed, the coil of the contactor KM3 is energized. The motor M4 starts to rotate forward, and the grinding wheel begins to rise. When it reaches the required position, release SB6, and KM3 coil will be de-energized and released. Motor M4 will stop, and the grinding wheel will stop rising.

Press the inching button SB7, the contactor KM4 coil is energized. The grinding wheel lifting motor M4 starts to rotate in reverse way, and the grinding wheel goes down. When it reaches the required position, release SB7, and the KM4 coil is de-energized and released. The motor M4 stops, and the grinding wheel stops descending.

In order to prevent the forward and reverse circuits of the motor M4 from being connected at the same time, the normally closed contacts of the contactors KM4 and KM3 are connected in series in the opposite circuit for interlocking control.

3) Analysis of the control circuit of the electromagnetic workbench

For the electromagnetic workbench, the electromagnetic chuck is a fixture to fix the workpiece. The magnetic field generated by the energized conductor in the iron core is used to attract the workpiece of ferromagnetic material for processing. Compared with mechanical clamps, electromagnetic chuck has the following advantages: rapid clamping and no damage to workpiece, and it supports several small workpieces clamped at a time, the workpiece have free expansion and contraction when it is heated. Therefore, the electromagnetic chuck is widely used on the surface grinder.

(1) The control circuit of the electromagnetic chuck includes three parts: a rectifier device, a control device and a protection device.

(2) The rectifier device is composed of a transformer TC, a single-phase bridge full-wave rectifier and supplies 120 V DC power.

The control device is composed of buttons SB8, SB9, SB10 and contactors KM5, KM6, etc.

① The magnetization processes. Press the magnetized button SB8, the KM5 coil of the contactor is energized, the main contacts of KM5 are closed. The YH coil of the electromagnetic chuck is energized, and the worktable is magnetized to attract the workpiece. At the same time, the self-locking contact is closed and the interlocking contact is disconnected.

After the grinding process is completed, remove the processed workpiece (first press SB9 to cut off the DC power supply of the electromagnetic chuck). Since the chuck and the workpiece have residual magnetism, the chuck and the workpiece need to be demagnetized.

② The demagnetization processes. Press the inching button SB10, the KM6 coil of the contactor is energized. The two main contacts of KM6 are closed, and the electromagnetic chuck is supplied with reversed-phase direct current to demagnetize the worktable and the workpiece. During demagnetization, in order to prevent the worktable from being reversely magnetized due to too long time and attracting the workpiece again, the contactor KM6 adopts inching control.

(3) The protection device consists of a discharge resistor R, a capacitor C, and a zero-voltage relay KA. The electromagnetic chuck is a large inductance, and a large amount of magnetic field energy is stored when magnetizing the workpiece. When it is disconnected from the power supply, the two ends of the suction cup generate a large self-induced electromotive force, which will damage the coil and other electrical appliances. Therefore, a discharge circuit is formed by a resistor and a capacitor. According to the characteristic that the voltage at both ends of the capacitor cannot change suddenly, the voltage change at both ends of the electromagnetic chuck coil tends to be slow, and the resistance is used to consume electromagnetic energy. If the parameters are properly selected, the circuit can form an attenuated oscillation circuit at this time to demagnetize, which would be very effective. During the processing, if the power supply voltage is insufficient, the electromagnetic chuck will not hold the workpiece firmly, which will cause the workpiece to be driven out by the grinding wheel and other serious accidents. Therefore, a zero-voltage relay KA is installed in the circuit. Connect its coil in parallel to the DC power supply, and its normally open contacts are connected in series in the control circuit of the hydraulic pump motor and the grinding wheel motor. If the electromagnetic chuck is not able to suck the workpiece firmly, the voltage relay KA will be released to make the hydraulic pump motor and the grinding wheel motor stop working to ensure safety.

4) Analysis of lighting and indicator circuit

EL in Figure 1-4 is a lighting lamp, its working voltage is 36 V, supplied by the transformer TC. QS2 is a lighting switch. HL1, HL2, HL3, HL4 and HL5 are indicator lights (HL6 and HL7 are not involved in this task), and their working voltage is 6.3 V, which is also supplied by the transformer TC. The functions of the five indicator lights are as follows.

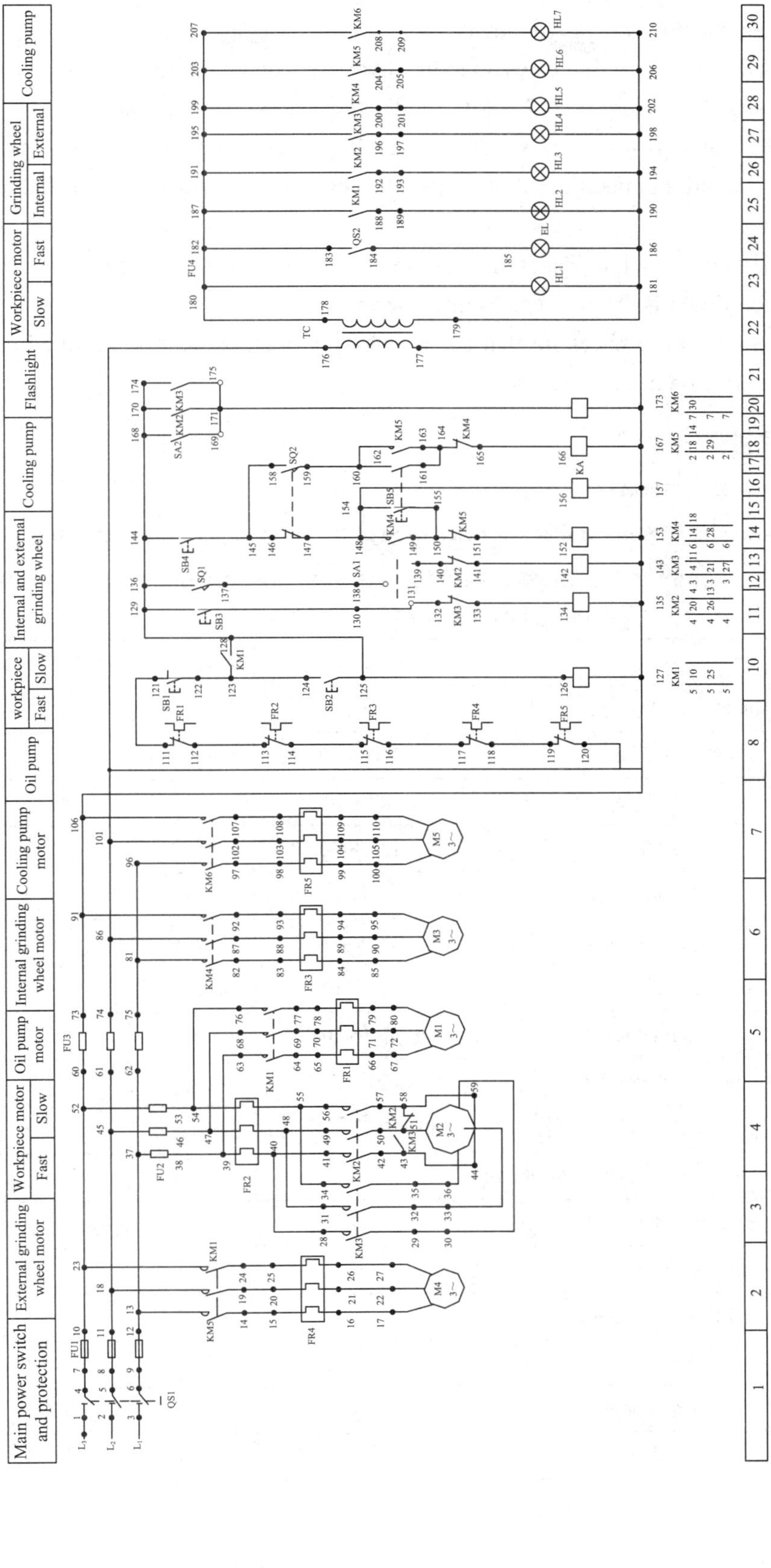

Figure 1-4 Electrical schematic diagram of M7120 surface grinder

(1) If HL1 is on, it indicates that the power supply of the control circuit is normal; if it is off, it indicates that the power supply is faulty.

(2) If HL2 is on, it indicates that the hydraulic pump motor M1 is running and the workbench is reciprocating; if HL2 is off, it indicates that M1 has been stopped.

(3) If HL3 and HL4 are on, it indicates that the grinding wheel motor M2 and the cooling pump motor M3 are in operation; if HL3 and HL4 are not on, it indicates that M2 and M3 have been stopped.

(4) If HL5 is on, it indicates that the grinding wheel lifting motor M4 is in the rising working state; if HL5 is off, it indicates that M4 has been stopped.

Please notice: after checking that the wiring is correct, it must be powered on and tested under the supervision of the teacher.

✍ Practice and Operation

1. Task preparations

Prepare the basic tools and accessories used in the training as shown in Table 1-4.

Table 1-4 Common troubleshooting tools and accessories for the electrical control circuit of the M7120 surface grinder

No.	Name	Model specification	Unit	Amount
1	Three-phase leakage switch	DZ47LE-32	Piece	1
2	1P fuse	RT14-20	Piece	6
3	2P fuse	RT8-30	Piece	1
4	3P fuse	RT18-30	Piece	1
5	Cross switch	LS1-1	Piece	1
6	Travel switch	LX19-001	Piece	4
7	AC contactor	CJ20-10	Piece	6
8	Thermal relay	JR36-20	Piece	3
9	Traction electromagnet	MQ1-127V	Piece	1
10	Transformer	150W 380V 130/110	Piece	1
11	Three-phase AC asynchronous motor		Set	4
12	Terminal board, wire duct, wire			Appropriate amount for each
13	Screwdriver		Piece	1
14	Electrician's pliers		Piece	1
15	Needle nose pliers		Piece	1
16	Wire stripper		Piece	1
17	Multimeter		Piece	1

2. Task procedures

Task procedures are as follows.

(1) Record the failure phenomena of the training unit of the control circuit of the M1720 surface grinder.

(2) Detect and record the open circuit and short circuit details of the control circuit of the M1720 surface grinder.

(3) Determine the fault location according to the following fault and symptom comparison details.

① Open circuit between 016–017: for the hydraulic pump motor, one phase is missing.

② Open circuit between 037–038: for both the grinding wheel motor and the cooling pump motor, one phase is missing (the same phase).

③ Open circuit between 039–040: for the grinding wheel motor, one phase is missing.

④ Open circuit between 048–062: for the grinding wheel descending motor, one phase is missing.

⑤ Open circuit between 061–068: the control transformer losts one phase and the control circuit fails.

⑥ Open circuit between 085–101: the control circuit fails.

⑦ Open circuit between 099–100: the hydraulic pump motor does not start.

⑧ Open circuit between 087–150: voltage KA relay does not operate, and hydraulic pump, grinding wheel cooling, grinding wheel lifting and electromagnetic chuck can not be started.

⑨ Open circuit between 120–121: the grinding wheel fails to rise.

⑩ Open circuit between 128–138: it represents magnetization and demagnetization failure of the electromagnetic chuck.

⑪ Open circuit between 136–137: the electromagnetic chuck cannot be magnetized.

⑫ Open circuit between 142–143: the electromagnetic chuck cannot be demagnetized.

⑬ Open circuit between 146–147: there is no direct current in the rectifier circuit, and the voltage KA does not operate.

⑭ Open circuit between 077–170: the lighting lamp does not turn on.

⑮ Open circuit between 159–164: it represents magnetization failure of electromagnetic chuck.

⑯ Open circuit between 174–175: the electromagnetic chuck cannot be demagnetized.

3. Precautions

(1) Establish groups and divide labor reasonably. The training is usually carried out in small groups, and each group consists of 2 to 3 people. The experiment usually involves wiring, load adjusting, meter reading, and data recording. The members of the experimental team should have a clear and reasonable division of labor, stand in their respective positions, coordinate their work, and conduct continuous consultations during the experiment to ensure that the experimental operation is standardized and coordinated.

(2) Select unit and meter. Before the training officially starts, it is necessary to select the appropriate unit and meter (including the range grade of the instrument) according to the training purpose, and record their models and parameters.

(3) Reading and exception handling. When reading the meter, pay attention to the magnification, minimum scale value, unit and significant digits. If an abnormality is found, immediately cut off the power supply, find the cause, and eliminate the fault.

4. Notes and Summary

____

____

____

____

____

____

【Extended Practices】

Exercise 1 Maintenance and troubleshooting of Z3050 radial drilling machine circuit

1. Training preparations

(1) Tools: screwdriver, electrician's pliers, wire stripper, needle-nose pliers, etc.

(2) Meter: 1 multimeter.

(3) Equipment: the required equipment is shown in Table 1-5.

Table 1-5 Equipment list required for Z3050 radial drilling machine circuit troubleshooting

No.	Name	Model specification	Amount
1	Three-phase leakage switch	DZ20Y-10	1
2	Three-phase leakage switch	DZ47-60	2
3	1P fuse	RT14-20	2
4	Universal switch	LW5-16/3	1
5	Cross switch	LS2-2	1
6	Travel switch	LX19K-B	2
7	AC contactor	CJ20-10	5
8	Thermal relay	JR36-20	2
9	Time Relay	JS7-2A	5

Continued table

No.	Name	Model specification	Amount
10	Transformer 1	150W　380V/127V/36V	1
11	Transformer 2	15W　220V/6V	1
12	Micro-motor	12V　6 revolutions	3
13	Three-phase AC asynchronous motor		4
14	Terminal board, wire duct, wire		Appropriate amount for each

2. Training equipment and content

1) Main circuit analysis

The electrical schematic diagram of Z3050 radial drilling machine is shown in Figure 1-5. There are four motor. Except that the cooling pump motor is directly started by a switch, the other three asynchronous motors are all started by contactors. The four motors are described below:

(1) M1 is the spindle motor, which is controlled by the AC contactor KM1. It only requires one-way rotation. The forward and reverse rotation of the spindle is operated by a mechanical handle. M1 is installed on the top of the spindle box to drive the spindle and feed transmission system. The thermal relay FR1 is an overload protection component. Short-circuit protection is an electromagnetic trip device in the main power switch.

(2) M2 is a rocker arm lifting motor, which is installed on the top of the main shaft, and uses contactors KM2 and KM3 to control forward and reverse rotation. Because the motor works for a short time, there is no overload protection appliance.

(3) M3 is a hydraulic oil pump motor, which can do forward and reverse rotation. The start and stop of forward and reverse rotation are controlled by contactors KM4 and KM5. The thermal relay FR2 is an overload protection device for the hydraulic oil pump motor. The main function of the M3 motor is to supply pressure oil to the clamping device to realize the clamping and loosening of the rocker arm and the column.

(4) M4 is a cooling pump motor with very low power, which is directly started and stopped by a switch.

2) Control circuit analysis

(1) Preparations before driving the drilling machine. For safety, the drilling machine has the function of “door open and power off at the same time”. Therefore, before driving the drilling machine, we should keep the electric door cover closed at the bottom of the column and the rear of the rocker arm to turn on the power. We need to keep the switch cover of the position switch closed and turn on the power switch QF1, then the power indicator HL1 is on, which indicates that the electrical circuit of the machine tool has entered a live state.

(2) Control of spindle motor M1. Press the start button SB3, and the contactor KM1 will be closed and self-locked, then the main motor M1 will start running. Press the stop button SB2 to release the contactor KM1, then the spindle motor M1 stops rotating.

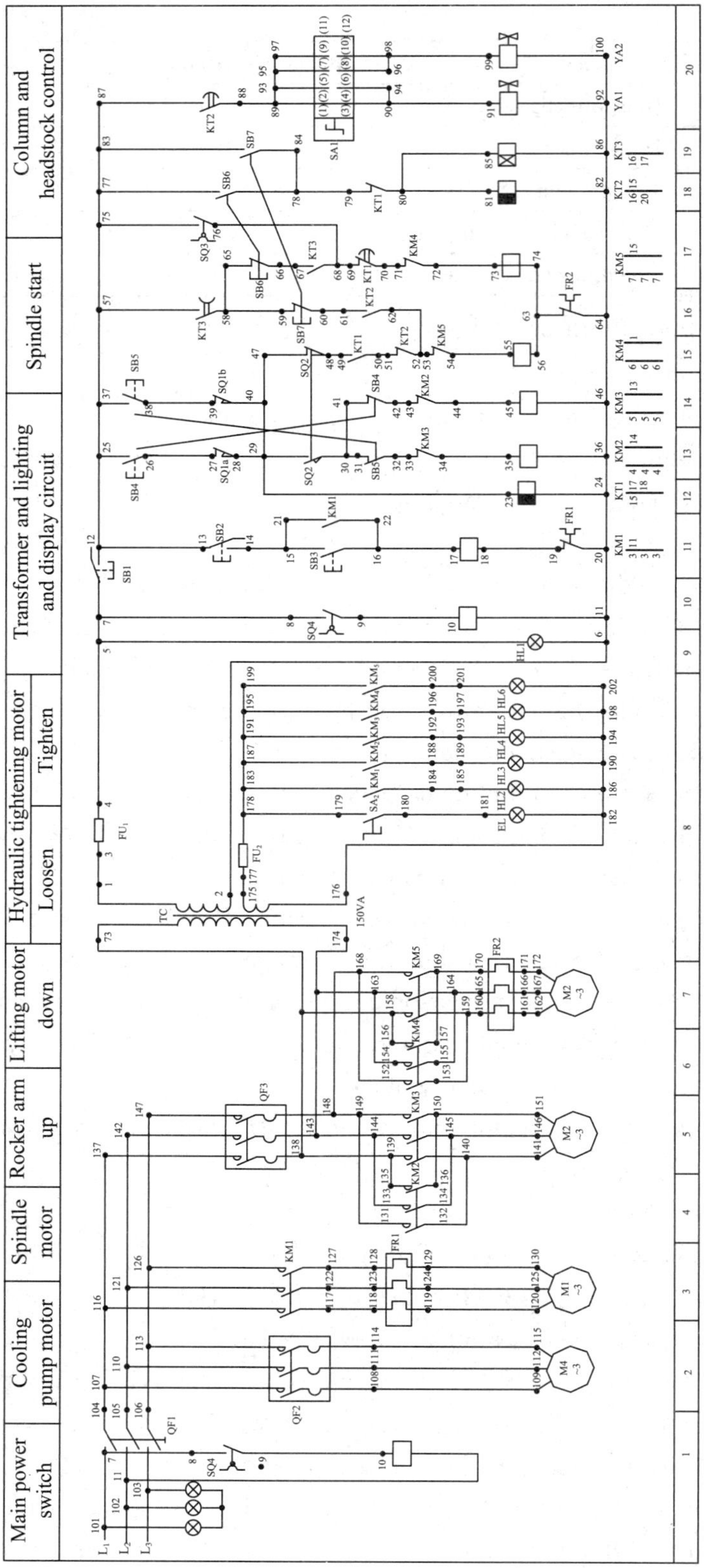

Figure 1-5　Electrical schematic diagram of Z3050 radial drilling machine

(3) Lifting control.

① Rocker arm up: Press the up button SB4, the time relay KT1 is energized to close, and its instantaneous moving-close contact is closed, the contactor KM4 coil is energized, and the hydraulic pump motor M3 starts to rotate in the forward direction to supply pressure oil. The pressure oil enters the "releasing oil chamber" of the rocker arm through the distribution valve to push the piston to move, and the piston pushes the diamond block to loosen the rocker arm. At the same time, the piston rod passes through the spring leaf position switch SQ2, so that the moving-break contact is disconnected, and the moving-close contact is closed. The moving-break contact cuts off the coil circuit of the contactor KM4, the main contact of KM4 is disconnected, and the hydraulic pump motor stops working. The moving-close contact is energized and the power is transferred to the coil of the AC contactor KM2, the main contact is connected to the power supply of M2, and the rocker lift motor starts to rotate forward. The rotation drives the rocker arm to rise. If the rocker arm is not released at this time, the normally open contact of the position switch SQ2 is not closed. The contactor KM2 cannot be closed, and the rocker arm cannot rise.

When the rocker arm rises to the desired position, release the button SB4. The contactor KM2 and the time relay KT1 will be de-energized and released at the same time. M2 will stop working, and the rocker arm will stop rising.

As the time relay KT1 is de-energized and released, after a time delay of 1 to 3 seconds, its time-delayed normally-closed disconnection contact is closed, so that the contactor KM5 is closed, and the hydraulic pump motor M3 rotates in the reverse direction. The internal pressure oil enters the "clamping oil chamber" of the rocker arm through the distribution valve, and the rocker arm is clamped. While the rocker arm is clamped, the piston rod disconnects the movable-break contact of the position switch SQ3 through the spring sheet, and the KM5 is de-energized and released, and finally the M3 stops working, the whole set of actions of loosening, rising and clamping of the rocker arm is completed.

② Rocker arm down. Press button SB5, the time relay KT1 is energized and closed, and its normally open contact is closed. The coil power of KM4 is turned on, and the hydraulic pump motor M3 starts to rotate forward to supply pressure oil. Similar to the process described above, the rocker arm is released first, and then the position switch SQ2 is pressed to open its normally closed contact, so that KM4 is de-energized and released, and the hydraulic pump motor stops working, and its normally open contact is closed to make KM3 energized, the rocker arm lifting motor M2 runs in the reverse direction, driving the rocker arm down.

When the rocker arm is lowered to the desired position, release the button SB5, and the contactor KM3 and the time relay KT1 are de-energized and released at the same time. M2 stops working, and the rocker arm stops descending.

As the time relay KT1 is de-energized and released, after a delay of 1 to 3 seconds, its time-delayed normally-closed disconnection contact is closed; the KM5 coil is energized; the hydraulic pump motor M3 rotates in the reverse direction, and the rocker arm is clamped. When the rocker arm is clamped, position SQ3 is disconnected, KM5 is powered off and released, and finally the motor M3 stopped, and the whole set of actions of loosening, lowering and clamping of the rocker arm is completed.

The position switches SQ1a and SQ1b are used to prevent the overtravel of the rocker arm. When the rocker arm rises to the limit position, SQ1a moves, the contactor KM2 is de-energized and released, M2 stops running, and the rocker arm stops rising; when the rocker arm drops to the limit position, SQ1b moves, the contactor KM3 is de-energized and released, and M2 stops rotating, the rocker arm stops descending. The automatic clamping of the rocker arm is controlled by the position switch SQ3.

(4) Clamping and loosening control of column and headstock. The loosening (or clamping) of the column and the spindle box can be carried out at the same time or separately, and it is controlled by the changeover switch SA1 and the button SB6 (or button SB7). SA1 has three positions: when it is pulled to the middle position, the column and headstock are released (or clamped) simultaneously; when it is pulled to the left position, the column is clamped (or loosened); when it is pulled to the right position, the headstock is clamped (or release). Button SB6 is the release button, and button SB7 is the clamping button.

When the column and the spindle box are loosened and clamped at the same time, move the transfer switch SA1 to the middle position, and then press the button SB6, and the time relays KT2 and KT3 will be energized at the same time. The normally open contact of KT2's time-delay disconnection is closed, and the electromagnets YA1 and YA2 are electrically attracted. The time-delayed normally-open disconnection contact of KT3 is closed after 1 to 3 seconds. After that, KM4 is closed and the hydraulic pump motor M3 rotates forward. The supplied pressure oil enters the column and the spindle box to release the oil cavity, then the column and the headstock are released at the same time.

The working principle of the simultaneous clamping of the column and the headstock is similar to that of loosening, just change button SB6 to button SB7, contactor KM4 to KM5, and M3 from positive to reverse.

If you want to control the headstock separately, you can move the switch SA1 to the right position, press the button SB6 (or the button SB7). At this time the coils of the time relay KT2 and KT3 are energized at the same time, and the solenoid YA2 is sucking and energized independently. The headstock can be loosened (or clamped) independently.

Release the button SB6 (or button SB7), the coils of the time relays KT2 and KT3 are de-energized and released. The normally-open contact of the KT3 is energized and closed momentarily, and the coil of the contactor KM4 (or KM5) is de-energized and released. The hydraulic pump motor stalls. After a delay of 1 to 3 seconds, the coil of electromagnet YA2 is de-energized and released, and the operation of unclamping (or clamping) the spindle box ends.

In the same way, pull the switch to the left to loosen or clamp the column individually.

3. Task procedures

(1) Record the fault phenomena of the Z3050 radial drilling machine control circuit training

unit.

(2) Detect and record the open circuit and short circuit details of the Z3050 radial drilling machine control circuit.

(3) Determine the fault location according to the fault phenomena comparison details.

4. Precautions

Refer to Task 3 in this module.

5. Notes and summary

Module 2 Comprehensive training of power electronic circuit technology

【General Introduction】

This module has designed three typical tasks for power electronic circuit assembly, adjustment and testing for senior electrician professionals and learners in related majors. The tasks are mainly based on principle recognition, circuit assembly and adjustment, and more complex circuit testing. The content includes comprehensive training of three-phase bridge-type fully-controlled rectifier circuit, comprehensive training of three-phase AC voltage regulating circuit, comprehensive training of PWM (Pulse Width Modulation) controlled switching stabilized power supply and some supplementary extension exercises. Through the training in this module, the trainees should acquire the knowledge and skills required by the electrician advanced power electronic circuit direction.

【Training Objectives】

1. Knowledge objectives

(1) Familiar with the structural characteristics of the three-phase bridge-type fully-controlled rectifier circuit, the connection method of the rectifier transformer and synchronous transformer.

(2) Master the application of KC785 integrated trigger circuit, and master the working principle and debugging method of three-phase thyristor integrated trigger circuit (including the test and analysis of the voltage waveform at each point).

(3) Familiar with the working principle of three-phase AC voltage regulating circuit.

(4) Familiar with the working principle of the switching regulated power supply composed of FET circuit.

(5) Learn to analyze the waveform of the switching regulated power supply.

2. Skill objectives

(1) Be able to analyze the voltage and current waveforms of the three-phase bridge-type fully-controlled rectifier circuit (when the resistance is the load) under different conduction angles.

(2) Master the wiring, setting and debugging of the three-phase AC voltage regulating circuit, and analyze the effective value and waveform of the voltage under different conduction

angles of the three-phase AC voltage regulating circuit (when the resistance is a load).

(3) Master the characteristics of PWM control, and the setting and adjustment of PWM integrated circuits.

3. Operating standards

The training aims to cultivate the application ability, more importantly, to improve the learners' self-learning ability, analytical ability, practical ability and innovation ability. The operation points include strict training for standard operation, circuit reading and analysis, system performance (or function) analysis and improvement, system debugging and operation, troubleshooting, etc.

As high-level training for power electronics and automatic control systems, the training aims not only to perform wiring of unit circuits, but to perform performance analysis (including data analysis and waveform analysis), improve system performance, and eliminate possible failures. The training operation is as follows:

(1) Establish groups and divide labor reasonably. The training is usually carried out in groups, each group consists of 2 to 3 people. The training usually involves wiring, load adjusting, meter reading, and data recording. The members of the training team must have a clear and reasonable division of labor, coordinate their work, and conduct continuous consultations during the training to ensure standardized and coordinated training operations.

(2) Select the unit and meter. Before wiring, select appropriate unit and meter (including the range of the instrument) according to the purpose of the training, and record their models and parameters.

(3) Wiring according to the diagram. The wiring method is first in series and then in parallel: from beginning to end, first connect in series to form a main circuit, and then connect the components one by one in parallel. In order for convenience, it is best to use the same color wire for each channel. After the wiring is completed, another member of the team must check that there is no error before the circuit is closed for training.

(4) Reading and exception handling. When reading the meter, pay attention to the magnification, minimum scale value, unit and significant digits. If an abnormality is found, immediately cut off the power supply, find the cause, and eliminate the fault.

After the training is completed, perform a preliminary analysis of the data to check whether there are any abnormalities. If there are any, the relevant data at this point will be reviewed again in the training. If any errors are found, they must be corrected.

(5) Training and stitching. After the training is completed, due to a relaxed mood, it is often the time when accidents are most likely to occur. Therefore, you must turn off the power before removing the cable, wait a little time to discharge the charge stored in the capacitor, and then remove the wire, and organize the instruments and training equipment.

4. Safety standards

In order to ensure personal safety and equipment safety during training, training members must strictly abide by the following safety operating procedures.

(1) The trainer should ensure the training device to have reliable grounding protection, leakage protection, overload protection, insulated table top, insulation carpet and safety patch cord.

(2) During the training, the human body should not come into contact with bare live objects.

(3) When wiring and disconnecting, it must be done with the power supply cut off.

(4) When close the circuit, all members of the training team must make clear and agree with the operation.

(5) When encountering abnormal situations, such as peculiar smell (such as the burning smell of varnish), overheated or excessive vibration unit (or running too much noise), burned out fuse, sparks in the electrical unit, over full scale in the meter (or reverse bias), the power should be cut off immediately to find the causes and eliminate the faults.

(6) When the DC motor starts directly, it must be ensured that the excitation power supply has been connected. If it is the armature power supply to be connected rather than the excitation power supply, it is easy to cause a flying accident. Secondly, a (starting) variable resistor should be connected to the armature circuit in series, and put the varistor slip point at the maximum resistance value. Otherwise, it will cause excessive current in the armature circuit (approximately short circuit) and burn out the motor.

(7) Before energize the unit, check whether the motor can rotate flexibly (check if the rotor is stuck). If it can’t rotate flexibly, you should re-adjust the rotor.

(8) The total power supply of the training room should be controlled by the training teacher. Other personnel can only operate with the teacher’s permission. It is not allowed to switch on without permission. If switch on rashly, serious accidents may occur.

(9) When using the resistor, note that the resistor cannot be set to zero (or a position with a small resistance value), as it may burn the varistor. During training, first adjust the high-value resistance, and then adjust the low-value resistance.

5. Quality Standards

Carry out analysis, research and discussion based on the measured data, and record the phenomena observed in the training. Students should process data and analyze graphics. It is a very important practices to summarize conclusions and write scientific and technological documents. Therefore, attention should be paid to write the training reports under the requirements of clear thinking, concise text, tidy diagrams, reasonable analysis, clear conclusions, and standardized and neat writing.

【Training Requirements】

1. Preparations of tools, instruments and equipment

The required components for Module 2 is listed as follows.

Table 2-1 List of components required for Module 2

No.	Name	Unit	Amount
1	Yalong YL-209 training device unit (13)	Piece	1
2	Dual trace oscilloscope	Set	1
3	Rheostat	Piece	1
4	Electric meter	Piece	1
5	Yalong YL-209 training device units (8), (9)	Piece	2
6	Yalong YL-209 training device unit (11)	Piece	1
7	Three incandescent lamps (connected to a three-phase delta load)	Piece	3
8	Multimeter	Piece	1

2. Knowledge assessment points

(1) The KC trigger circuit and its working principle.

(2) The principle and application of three-phase half-wave controllable rectifier circuit.

3. Skills assessment points

(1) The working principle of three-phase controllable rectifier circuit.

(2) The types and working principles of thyristor trigger circuit.

(3) The types and characteristics of three-phase controllable rectifier circuit.

【Work Tasks】

Task 1 Comprehensive training of three-phase bridge-type fully-controlled rectifier circuit

✍ Relevant knowledge

1. Basic components

1) Rectifier diode

The basic structure and working principle of rectifier diodes are the same as the diodes in electronic circuits. They are all based on semiconductor PN junctions to achieve forward conduction and reverse cut-off functions. Rectifier diode is uncontrollable device. Its turn-on and turn-off are completely determined by the voltage and current it bears in the main circuit. Common parameters of rectifier diodes are as follows:

(1) Maximum average rectified current I_F: it is the maximum forward average current allowed to pass through the diode during long-term operation. The current is determined by the junction area of the PN junction and the heat dissipation conditions. It should be noted that the average current through the diode cannot be greater than I_F value, and the heat dissipation conditions should be met. For example, the I_F of the 1N4000 series diode is 1A.

(2) Maximum reverse working voltage U_R: it is the maximum reverse voltage allowed to be applied across the diode. If the reverse working voltage is greater than this value, the reverse current (I_R) will increase sharply, and the unidirectional conductivity of the diode will be destroyed, causing reverse breakdown. Usually take half of the reverse breakdown voltage (U_B) as the maximum reverse voltage (U_R). For example, the U_R of 1N4001 is 50V, and the U_R of 1N4007 is 1000V.

(3) Maximum reverse current I_R: it is the reverse current allowed to flow through the diode under the maximum reverse working voltage. This current reflects the quality of the unidirectional conductivity of the diode. Therefore, the smaller the maximum reverse current value, the better the diode quality.

(4) Breakdown voltage U_R: it refers to the voltage value at the sharp bend point in the reverse volt-ampere characteristic curve of the diode. When the reverse of the diodes is in soft characteristic, it refers to the voltage value under a given reverse leakage current condition.

(5) Reverse recovery time T_{RE}: it refers to the recovery time under the specified load, forward current and maximum transient voltage.

2) IGBT

IGBT refers to an insulated gate bipolar transistor. It is a composite fully controlled voltage-driven power semiconductor device composed of GTR (Giant Transistor) and MOSFET (Metal Oxide Semiconductor Field Effect Transistor), which has two advantages: high input impedance of MOSFET and low conduction pressure drop of GRT. The saturation voltage of GTR is reduced, the current-carrying density is high, but the driving current is large; the MOSFET driving power is small, the switching speed is fast, but the conduction voltage drop is large, and the current-carrying density is small. IGBT combines the advantages of the above two devices, with low driving power and reduced saturation voltage. It is very suitable to be used in the fields of converter systems with DC voltage of 600V and above, such as AC motors and inverters.

The switching function of IGBT is described as follows: IGBT forms a channel by applying a positive gate voltage to provide a base current to the PNP transistor, then IGBT is turn-on. On the contrary, the reverse gate voltage is applied to eliminate the channel, cut off the base current, then IGBT is turn-off.

IGBT parameters are as follows:

(1) Maximum collector-emitter voltage U_{CES}. It is determined by the breakdown voltage that the PNP transistor inside the device can withstand.

(2) The maximum collector currents. It includes the rated DC current I_C and the maximum current with a pulse width of 1ms.

(3) Maximum collector power consumption P_{CM}. It refers to the allowable maximum dissipation power at normal operating temperature.

2. Three-phase bridge-type fully-controlled rectifier circuit

The three-phase bridge fully-controlled rectifier circuit is shown in Figure 2-1.

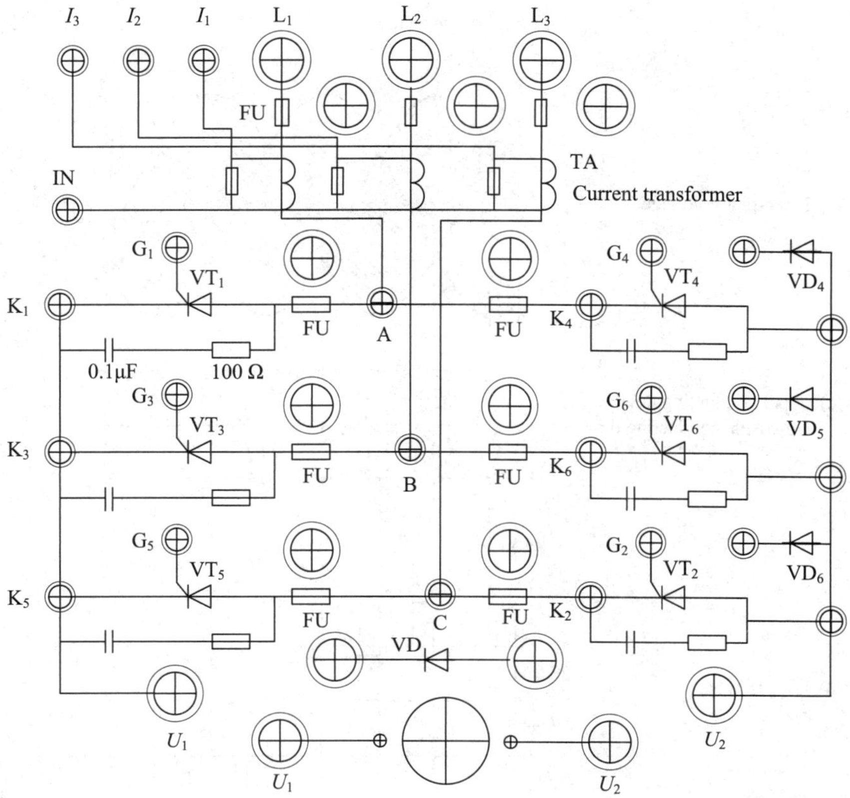

Figure 2-1 Three-phase bridge-type fully-controlled rectifier circuit

The turn-on sequence and characteristics of the 6 thyristors in the Figure 2-1 are as follows.

(1) The starting conduction point of thyristors(the natural commutation point), that is, the common cathode including VT_1, VT_3, VT_5 is the intersection of the three positive half waves of the three-phase voltage U_A, U_B and U_C; and the common anode including VT_4, VT_6, VT_2 is the intersection of the negative half-wave of the three-phase voltage U_A, U_B and U_C.

(2) In each of the common anode and common cathode thyristors, only one thyristor is connected, can a path form, such as path 1-2, 2-3, 3-4, 4-5, 5-6, 6-1, etc., see Figure 2-2. In this way, Thyristor turn-on sequence is $VT_1 \to VT_2 \to VT_3 \to VT_4 \to VT_5 \to VT_6$, and the interval pulse is 60°.

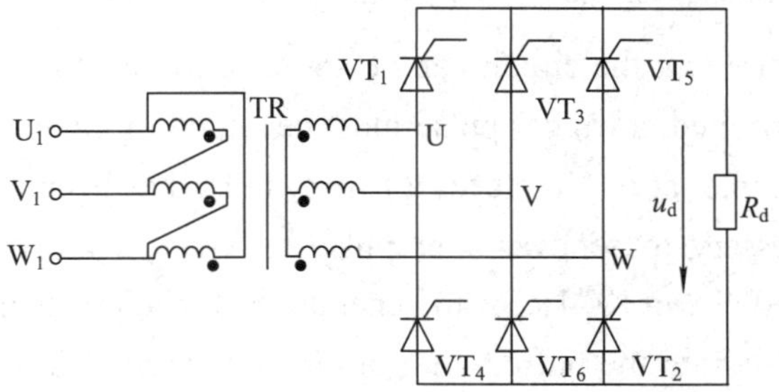

(a) Three-phase bridge-type fully-controlled rectifier circuit I

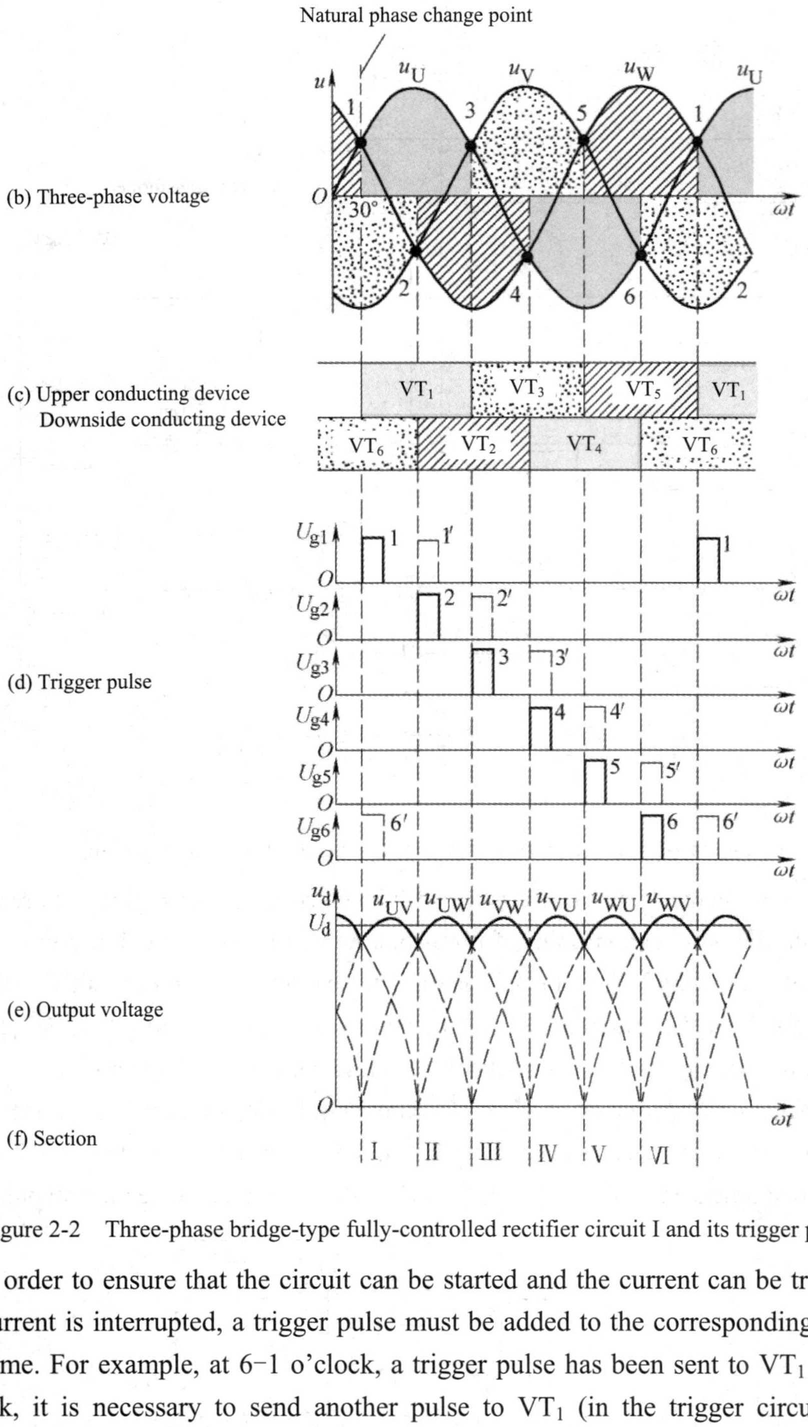

Figure 2-2　Three-phase bridge-type fully-controlled rectifier circuit I and its trigger pulse

(3) In order to ensure that the circuit can be started and the current can be triggered again after the current is interrupted, a trigger pulse must be added to the corresponding two tubes at the same time. For example, at 6−1 o'clock, a trigger pulse has been sent to VT_1 ahead, but at 1−2 o'clock, it is necessary to send another pulse to VT_1 (in the trigger circuit below, the function of the integrated circuit KC41C is to generate the complementary pulse), so each tube is triggered by double pulse with the pulse width of 60° see figure 2-2(b)(of course, a wide pulse with a pulse width greater than 60° can also be used, but the power consumption is large).

In Figure 2-1, TA is a current transformer (three phases in total), (HG1 type, 5 A/2.5 mA, load resistance is less than 100 Ω). Since the secondary side of the current transformer cannot be opened (open circuit will cause high voltage), so there is a load resistor on the secondary side.

3. Wiring of rectifier transformer and synchronous transformer

(1) The a rectifier transformer is mainly used to make the rectified output voltage and the working voltage of the motor appropriate. Since the working voltage of motor in this training is 110V, the three-phase bridge fully-controlled rectifier circuit voltage formula is U_d=2.34U_2 (U_d is the DC output voltage. U_2 is the phase voltage of the transformer secondary side). If U_d=110V is substituted into the above formula, U_2≈47 V.

(2) The rectifier transformer is connected in Dy type (Δ/Y type), which can effectively suppress the adverse effects of the third harmonic generated during rectification on the power grid. Here connect it as Dy11 (Δ/Y-11).

(3) In addition, the rectifier transformer plays a role of isolation, which is conducive to personal safety.

(4) The trigger circuit adopts the integrated trigger circuit KC785 whose synchronous voltage is a sawtooth wave. Since the synchronous voltage has to pass through the resistance-capacitance filter circuit, it will cause a phase lag (60°−70°), which needs to be compensated. Because the voltage zero-crossing point 30° ahead of the natural commutation point, the synchronous voltage 30° ahead of the main circuit voltage is available. Therefore, the connection method of Yy10 (Y/Y-10) is adopted.

The connection diagram of the synchronous transformer and rectifier transformer is shown in Figure 2-3.

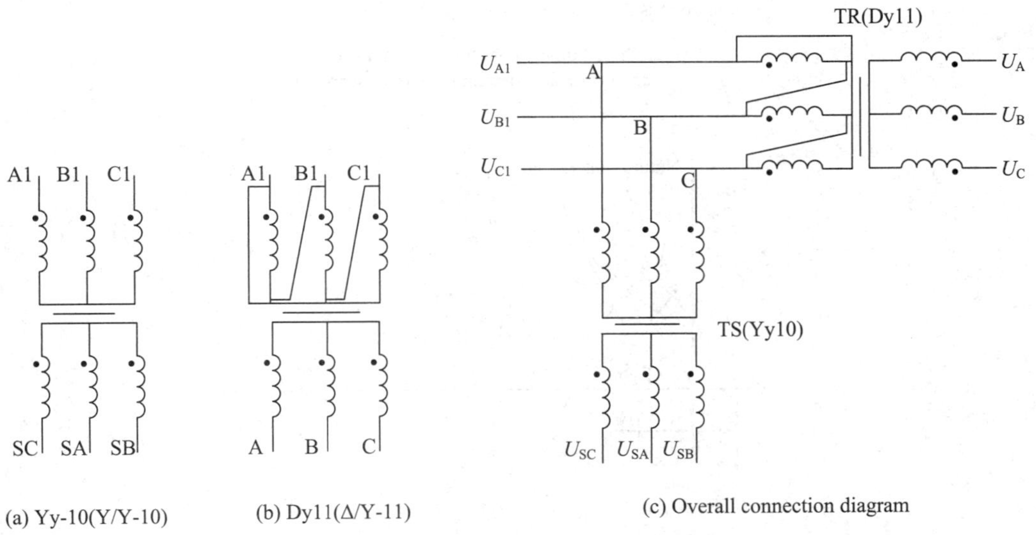

Figure 2-3 Connection diagram of synchronous transformer and rectifier transformer

Note: The clock representation in three-phase transformer connection: the clock long-hand (long needle) represents primary side phase voltage of transformer, the clock short-hand (short needle) represents secondary side phase voltage.

See from Figure 2-3(c), U_A on the secondary side of the rectifier transformer corresponds to U_{AB1} on the primary side, and U_{AB1} is 30° ahead of U_{A1}, so U_A (short needle) and U_{A1}(long needle) form 11 o'clock, see Figure 2-4(a).

It can also be seen from Figure 2-3(c) that U_{SA} corresponds to $-U_{B1}$, so that it can be seen from Figure 2-4 that U_{SA} is 30° ahead of U_A. Now after the capacitive-resistance filter circuit, the phase lags about 70°, so the voltage after the phase shift will lag U_A by about 40° (70°−30°). It lags behind the natural commutation point only by about 10° (40°−30°). This means that the phase shift range of the control angle α is 10° to 120°. The reason for not making the control angle start from 0° is to prevent the output voltage from being too high, and to make the phase shift range in the linear segment of the sawtooth wave.

In the Figure 2-4(b), U_{A1} is 220V, U_A is 47V, and U_{SA} is 16.5V.

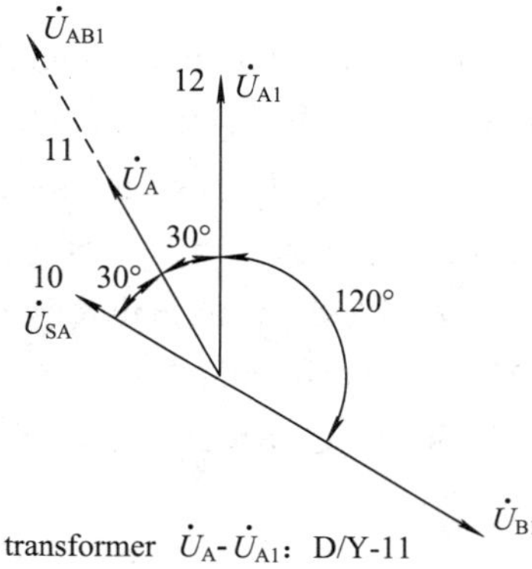

Rectifier transformer $\dot{U}_A$- $\dot{U}_{A1}$: D/Y-11

Synchronous transformer $\dot{U}_{SA}$- $\dot{U}_{A1}$: Y/Y-10

(a) Phase voltage phhasor diagram of rectifier transformer and synchronous transformer A

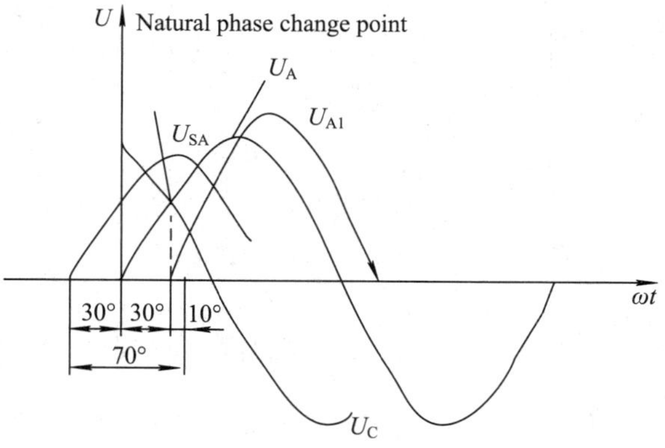

(b) The Synchronous voltage that lags behind the natural commutation 10°

Figure 2-4 Transformer phase relationship diagram

4. Three-phase thyristor integrated trigger circuit (KC trigger)

The core part of the three-phase thyristor integrated trigger circuit (as shown in Figure 2-5) is a circuit composed of three integrated trigger circuits N1, N2 and N3, which are TCA785 (domestic KJ785 or KC785) integrated circuits.

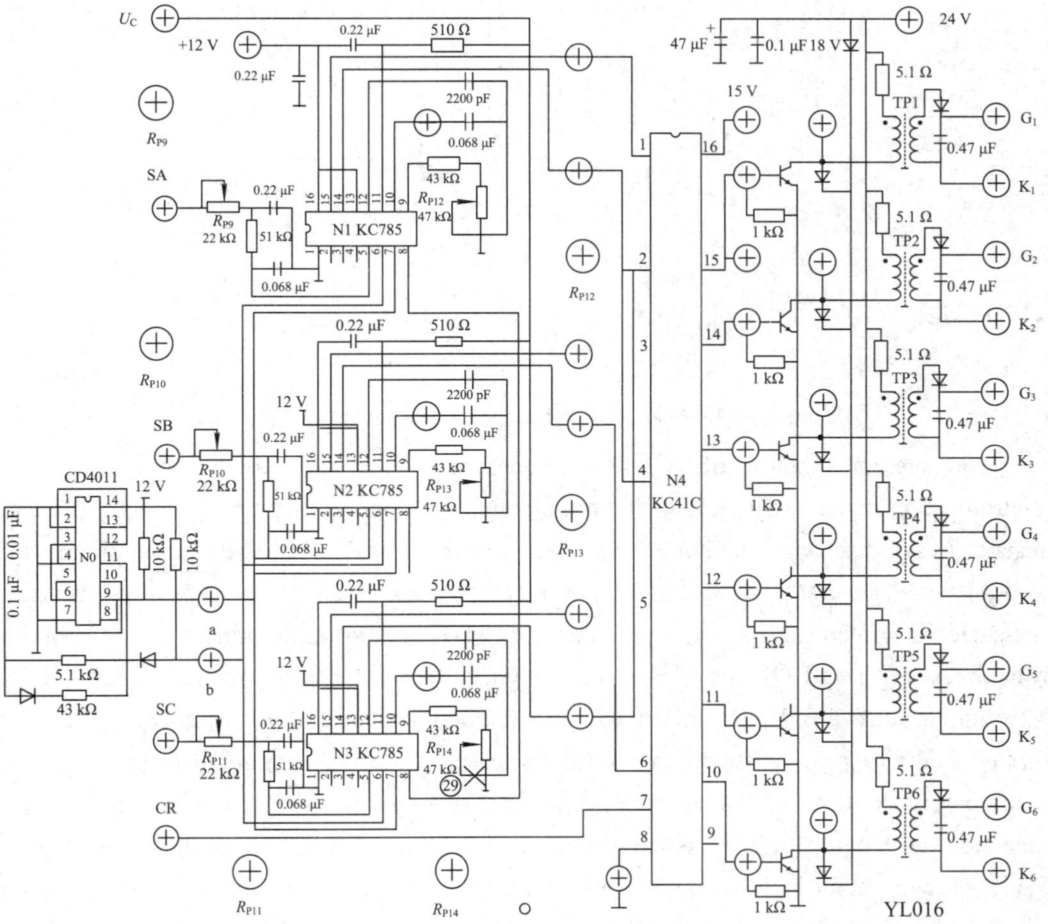

Figure 2-5 Three-phase thyristor integrated trigger circuit

The components of the three-phase thyristor integrated trigger circuit are as follow.

(1) TCA 785 is a third-generation thyristor monolithic phase-shift integrated trigger circuit developed by Siemens. Its output and input are compatible with CMOS(Complementary MOS) and TTL(Transistor-Transistor-Logic) levels. It has a wide voltage range and a large load driving capability. Each path can directly output 250 mA drive current. In its circuit structure, the sawtooth voltage range is wider, and it is more adaptable to ambient temperature. The operating power supply voltage range of this integrated circuit is −0.5−18 V. The schematic diagram of the internal structure of TCA785 is shown in Figure 2-6.

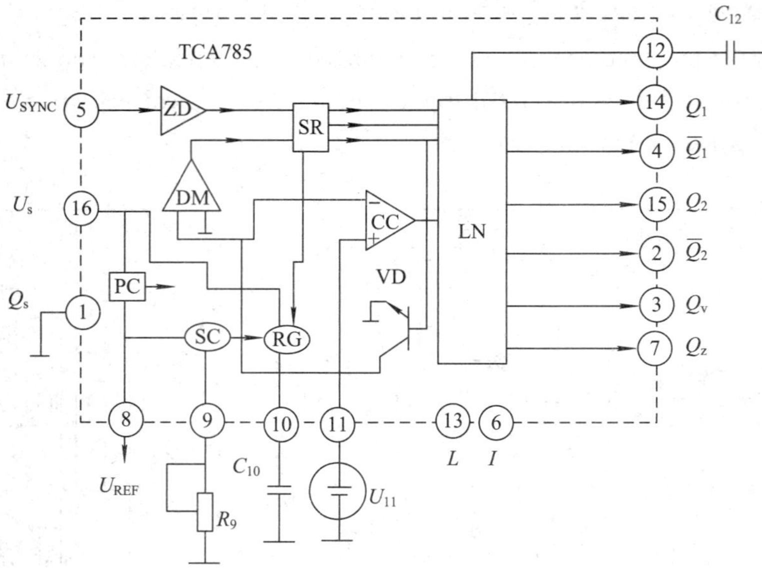

Figure 2-6　Schematic diagram of the internal structure of TCA785

The internal structure of TCA785 includes nine units: zero point discriminator (ZD), synchronization register (SR), constant current source (SC), control comparator (CC), discharge transistor (VD), discharge monitor (DM), level conversion and voltage regulation circuit (PC), sawtooth wave generator (RG) and output logic network. TCA785 is a dual-in-line 16-pin large-scale integrated circuit, each pin function: ⑯ (U_s)—power terminal; ① (Q_s)—ground terminal; ④ ($\bar{Q}_1$) and ② ($\bar{Q}_2$)—they output pulse 1 and 2 non-terminals; ⑭ (Q_1) and ⑮ (Q_2)—output pulse 1 and 2 terminals; ⑬ (L)—the output pulse Q_1, Q_2 width control terminal; ⑫ (C_{12})—output Q_1, Q_2 pulse width control terminal; ⑪ (U_{11})—Output pulse $\bar{Q}_1$, $\bar{Q}_2$ or Q_1, Q_2 phase shift control DC voltage input terminal; ⑩(C_{10})—external sawtooth wave capacitor connection terminal; ⑨(R_9)—sawtooth wave resistance connection terminal; ⑧(U_{REF})—TCA785's own output high Stable reference voltage terminal; ⑦(Q_z) and ③(Q_v)—two output logic pulse signal terminals by TCA785; ⑥(I)—pulse signal prohibition terminal; ⑤(U_{SYNC})—synchronous voltage input terminal.

The working process of TCA785 is described as follows: the synchronous voltage from the synchronous voltage source is sent to the zero point discriminator ZD after passing through a high-resistance resistor, after ZD detects its voltage zero crossing point, it is sent to the synchronous register for registration. The zero register signal in the synchronization register controls the generation of the sawtooth wave. The resistor R_9 determines the voltage charged by the constant current source SC. When the sawtooth voltage across the capacitor C_{10} is greater than the phase shift control when the voltage is U_{11}, a pulse signal is generated and sent to the output logic unit, as shown in Figure 2-7. It can be seen that the phase shift of the trigger pulse is controlled by the magnitude of the phase shift control voltage U_{11}, so the trigger pulse can be phase shifted in the range of 0° to 180°. For each half cycle, a narrow pulse with a width of about 30 μs appears at the output terminals Q_1 and Q_2. The pulse width can be determined by the 12-pin capacitor C_{12}. If pin 12 is grounded, the output pulse interval between Q_1 and Q_2 is 180°.

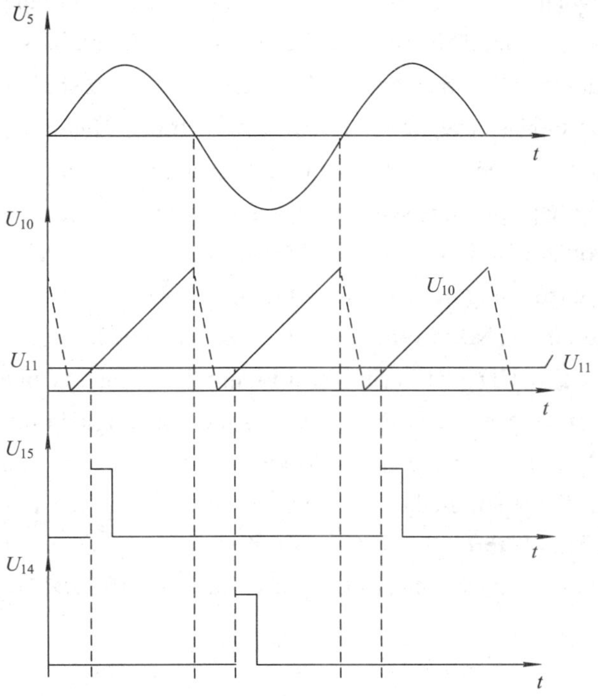

Figure 2-7 Generation of trigger pulse

The main technical data of domestic KC785 are as follows.

① Power supply voltage: DC +15V (allowable working range 12–18 V).

② Power supply current: no more than 10 mA.

③ Synchronous input terminal: it allows the maximum synchronization current 200 μA.

④ Phase shift voltage range: −0.5–(U_S−2)V(maximum value 13 V).

⑤ Phase shift range: no less than 170°.

⑥ Sawtooth wave amplitude: (U_S−2) V (maximum value 13 V).

⑦ Output pulse:

Amplitude: high level is no less than (U_S−2.5) V; low level is no more than 2 V.

Width: about 30 μs (without C12) (400–600) μs/nF (with C12).

Maximum output capacity: 55mA (pulse current flowing out).

2#3#4#7#Pulse voltage output terminal output capacity: no more than 2 mA (injection pulse current)

⑧ Package: 16-pin plastic dual in-line package.

⑨ Allowable operating temperature: −10–70℃.

(2) R_{P11}, R_{P12} and R_{P13} in Figure 2-5 are variable resistors at the N1, N2 and N3 pins. They

are used to adjust the slope of the three-phase sawtooth wave.

(3) SA, SB, SC in Figure 2-5 are three-phase synchronous voltages, which are introduced by the three terminals of synchronous transformers U_{SA}, U_{SB} and U_{SC}. After the phase lag is about 40° through the resistance-capacitance filter circuit, they are sent to 5* pin of N1, N2 and N3.

Due to the resistance and capacitance value error, the phase shift angle will be different, which will make the three-phase trigger waveform asymmetrical. Therefore, a variable resistor (22 kΩ) is added at the input of the synchronous voltage of each phase to adjust the phase shift, so that the three-phase output voltage phase is symmetrical and mutually different by 120°.

(4) The control voltage U_C is sent to the 11* pin of N1, N2 and N3 via the current limiting resistor at the same time (to compare with the sawtooth wave).

(5) The integrated circuit N0 in Figure 2-5 is CD4011, which includes four 2-input NAND gates. The circuit formed by CD4011 is shown in Figure 2-8, which is an externally excited ("0" effective) ring oscillation device. This circuit receives the pulse signal output by KC785 from the 7* pin of N1, N2 and N3, and after the circuit forms an oscillation, the pulse output by KC785 is made into a pulse train (the leading edge of the pulse train is steep, the amplitude is high, and the power consumption is small) through the 6* pin (which is the pulse signal prohibition terminal of N1, N2 and N3). The frequency range of the pulse train is 5–10 kHz.

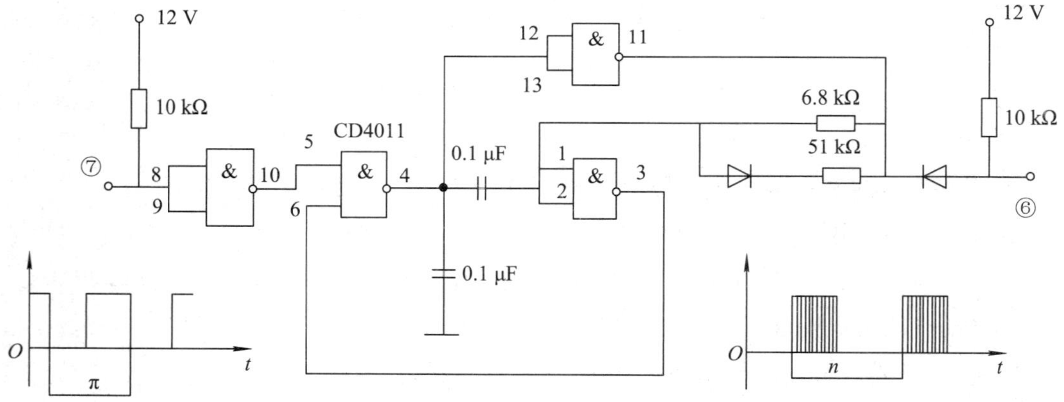

Figure 2-8　Ring oscillator

(6) N4 in Figure 2-5 is KC41C, and its internal structure schematic diagram and application examples are shown in Figure 2-9. Its function is to generate a supplementary pulse through the input diode for the basic pulse output from N1, N2 and N3 through 14* and 15* pins. For example, when the 2* pin inputs a pulse, it sends a pulse signal to the base of VT through the diode VD at the same time, so that VT_2 and VT_1 can be turned on at the same time. See Figure 2-9(a), VT_7 is an electronic switch. When the 7* pin is "0", VT_7 is cut off, and each channel will have pulse output; when the 7* pin is "1" (floating), the +15V voltage input from the 16* pin, VT_7 will be turned on and the output path will be blocked (set to zero). Therefore, lead 7* out as a blocking signal CR. The zener tube in the component provides a threshold voltage to prevent

false triggering. The 16* pin of the component is connected to the +15V power supply, and the 8* pin is grounded.

(7) The trigger pulse output by KC41C is amplified by power, and then sent to the G and K poles of the six thyristors VT_1–VT_6 through the pulse transformer.

(8) In Figure 2-5, in the freewheeling (diode) circuit on the primary side of the pulse transformer, an 18V voltage regulator tube is connected in series to reduce the pulse current rapidly (to increase the steepness of the pulse trailing edge). And the overvoltage should not be too large (<18V).

(9) The power supply of this circuit should be three groups of +12V, +15V and +24V, Please don't make mistakes.

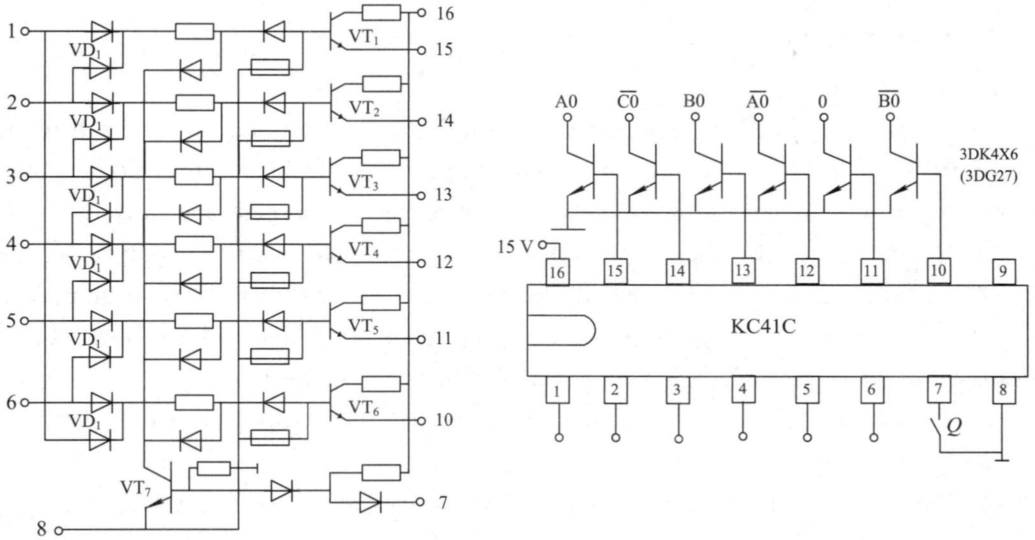

(a) KC41C internal schematic diagram (b) Schematic diagram of KC41C application example

Figure 2-9 KC41C internal structure schematic diagram and application examples

✍ Practice and operation

1. Task preparations

There are two steps in task preparation:

(1) Preview the "relevant knowledge" part of this module.

(2) Prepare the "three-phase bridge-type fully-controlled rectifier circuit" module, multimeter, and oscilloscope. The specific required device models are shown in Figure 2-5 and Table 2-1.

2. Training content

Training content and requirements are as follows.

(1) Understand the working principle of the three-phase bridge fully controlled rectifier circuit.

(2) Master the transformer connection group and star-delta connection method.

(3) Master the basic working principle of KC trigger.

(4) Operate according to the task procedures, write down the average value and waveform of the output voltage U_{SA} of the power supply U_{A1}, the rectifier transformer U_A, and the synchronous transformer, as well as the phase difference between them; write down the amplitude of the trigger pulse, waveform and phase of the VT_1−VT_6 in Figure 2-9; write down the values and waveforms of the voltage when the resistance load is at α=30°, α=60° and α=90°, decide whether their average value is consistent with the calculated value; When α=60°, decide the voltage waveform between K end and A end of VT_1.

Consider if VT_2 is damaged, what is the voltage waveform between A end and K end? Analyze the waveform, point out the normal and abnormal places, and analyze the causes.

3. Task procedures

The task procedures are as follows.

(1) Connect the rectifier transformer to the Dy11 connection, and the synchronous transformer to the Yy10 connection, without connecting the load. Connect their primary side to 220 V/380 V power supply. Use an oscilloscope to measure the amplitude and waveform of U_{A1}, U_A and U_{SA}, and observe whether the waveform when there is 380 V power supply is 30° ahead of the 220 V. At the same time, measure whether the ±12 V power supply voltage is normal.

(2) Cut off the power supply and connect the output U_A, U_B, and U_C of the rectifier transformer to the L_1, L_2 and L_3 input terminals of the main circuit respectively.

(3) A resistance load (varistor) is indirectly connected to the output terminals U_1 and U_2 of the main circuit.

(4) The trigger circuit is connected to +12 V, +15 V and +24 V power supplies, and the synchronous voltage (16.5 V) is input. The control voltage U_C terminal is connected to the regulated power supply. U_C is adjusted between 0−8 V, and U_C is adjusted around 4V first. Use a multimeter and an oscilloscope to observe the amplitude and waveform of the N1 pin 10* (sawtooth wave) and the output of 14* and 15* pins (dual pulse train).

It can be seen from Figure 2-7 that when the control voltage U_C (that is, U_{11} in the Figure 2-7) is the minimum, α is the minimum, and the output voltage is the maximum at this time. On the contrary, when U_C≈8V, the trigger pulse disappears and U_d=0.

Adjust R_{P11} so that the amplitude of the sawtooth wave of N1 is 7.8−7.9 V. When U_{C1} increases to the maximum (about 8V), adjust R_{P11} appropriately till the pulse of N1 disappears.

(5) By using the sawtooth wave of N1 as the reference, adjust R_{P12} and R_{P13} so that the slope of the sawtooth wave of N2 and N3 is the same as that of N1 (observed with an oscilloscope).

(6) Adjust the control voltage U_C to make U_C change from 0 V to 8 V, and observe the phase shift range of the pulse. Measure the 6 trigger pulses to check if the difference is 60°, and record the waveform of the trigger pulse.

(7) Measure the amplitude and phase of the output pulse of the 10^*-15^* pins of N4. If the trigger pulses are correct (as shown in Figure 2-2), after cut off the power supply, connect the output of the pulse transformer to the G and K poles of the corresponding six thyristors.

(8) Turn on the power, observe the value and waveform of the voltage on the resistive load, adjust the size of U_C so that the control angle α is 30°, 60°, 90° and 120° respectively, and record the average value and waveform of the voltage.

(9) Adjust the rheostat and U_C so that the current I_d=1.5A, and measure the voltage value output by the current transformer. (Between I_1 and I_2 or between I_2 and I_3).

(10) Measure the voltage waveform between K end and A end of VT_1 when α=60°.

(11) If one of the 6 thyristors (set VT_2 damaged-remove its trigger pulse) re-measure the amplitude and waveform of U_d to judge if the component is normal from the waveform of the thyristor.

4. Precautions

(1) Before the training, review the relevant basic knowledge of the power electronics related course, and carefully read the specific task procedures.

(2) Since there are many wiring modules in training, you should connect one unit, check one unit, and test whether it is normal. Only when it is ensured that the units are working properly, can the units be connected.

(3) There are many steps in the training that need to be compared and measured with an oscilloscope, and attention should be paid to find out the wiring of the common end of the two probes, otherwise it is easy to cause a short circuit.

5. Notes and summary

____

____

____

____

____

____

Task 2 Comprehensive training of three-phase AC voltage regulating circuit

✍ Relevant knowledge

The three-phase AC voltage regulating circuit is shown in Figure 2-10.

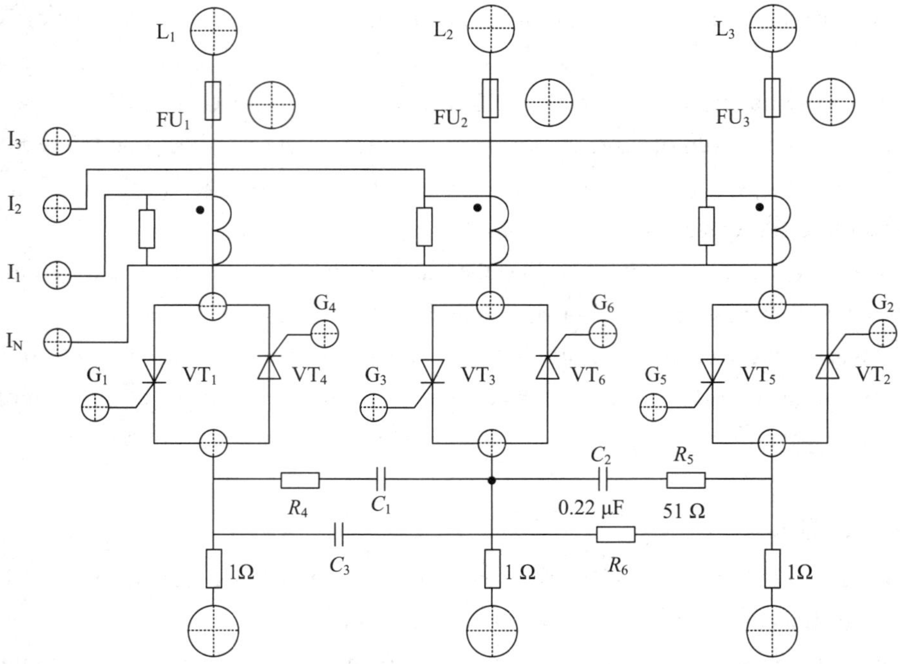

Figure 2-10 Three-phase AC voltage regulating circuit

The differences between the circuit shown in Figure 2-10 and the three-phase bridge fully-controlled rectifier circuit shown in Figure 2-1 are as follows.

① Each phase has two thyristors, which are powered in anti-parallel.

② The natural commutation point is the zero-crossing point of the voltage.

③ Circuit of Figure 2-10 is three-phase parallel output.

④ When the control angle α is small, three thyristors are in conduction at the same time; when α is large, two thyristors are in conduction at the same time.

⑤ The output voltage waveforms are different.

Let's describe the situation when α=0° and α=60°.

1. Control angle α = 0°

Different from the three-phase rectifier circuit, when α=0°, the trigger pulse is applied to the tube at the zero crossing of the corresponding voltage of each phase. This is equivalent to treating the thyristor as a diode. At this time, the three phases have forward and reverse currents. It is equivalent to a general three-phase AC circuit, and the current of each phase is $i_{\Phi}=\dfrac{u_{z\Phi}}{R}$.

The thyristor conduction sequence is $VT_1 \to VT_2 \to VT_3 \to VT_4 \to VT_5 \to VT_6$, the pulse interval is 60°. The conduction angle of each tube is θ=180°. Except for the natural change point, 3 of all thyristors are turned on at any time.

2. Control angle α=60°

When the control angle α=60°, the thyristor is in conduction situation, current waveform is the U-phase (see Figure 2-11). At the moment of ωt_1, VT_1 is triggered. A current loop is formed between VT_1 and VT_6. At this time, under the action of the online voltage u_{UV}, for the star-delta connection load, $i_U=\frac{u_{UV}}{2R}$. At the moment of ωt_2, VT_2 is triggered, the load voltage is u_{UW}, and the U-phase current is $i_U=\frac{u_{UV}}{2R}$ at this time. At the moment of ωt_3, VT_3 is triggered, VT_1 is turned off, and VT_4 is not turned on yet, so i_U=0. At the moment of ωt_4, VT_4 is triggered and turned on. Under the action of u_{UV} voltage, i_U forms a loop through VT_3 and VT_4. Similarly, during the period of ωt_5–ωt_6, u_{UW} forms a loop through VT_4 and VT_5. The i_U current waveform is shown by the hatched line in the Figure 2-11.

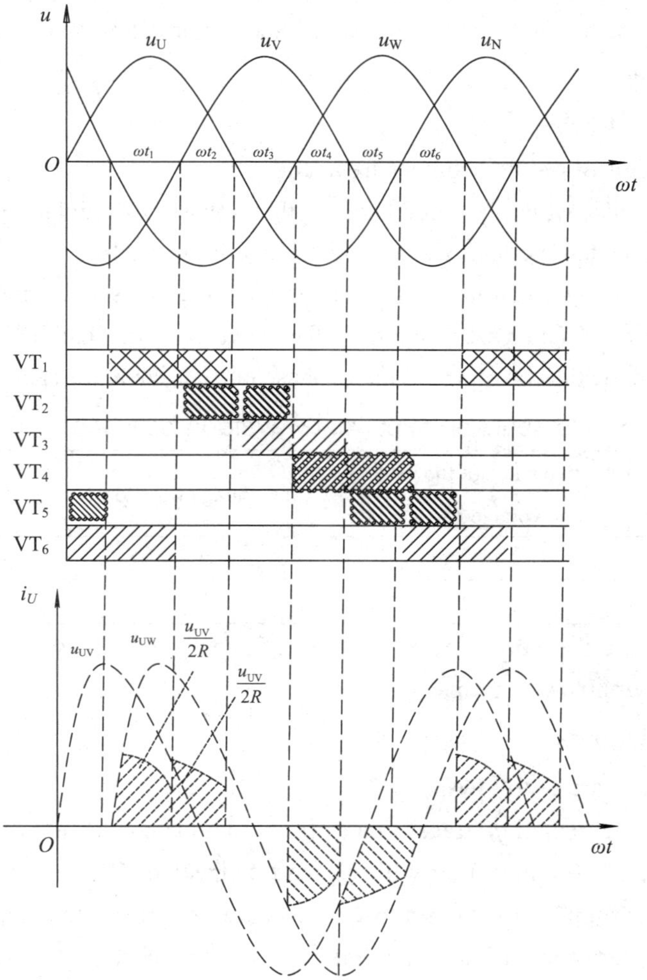

图 2-11 The thyristor conduction situation and the U-phase current waveform when α=60°

The i_V and i_W waveforms can be obtained by the same analysis.

Note:

(1) The trigger circuit is the same as Figure 2-5, and its working principle has been explained in Task 1 in this module.

(2) Isolation transformer and synchronous transformer comparison. To ensure safety, power is also supplied through an isolation transformer here. The connection diagram of isolation transformer and the synchronous transformer is the same as Figure 2-3. Its working principle has been explained in "Relevant Knowledge" in Task 1 in this module.

✍ Practice and operation

1. Task preparations

There are two requirements in task preparation.

(1) Clarify the working principle of the three-phase voltage regulating circuit.

(2) Prepare the components in the three-phase AC voltage regulating circuit, including oscilloscope, multimeter and connecting wire. See Figure 2-10 and Table 2-1 for specific required components.

2. Training content

The requirements of training content are as follows.

(1) Grasp the basic working principle of the three-phase AC regulating circuit.

(2) Complete the basic circuit wiring according to Figure 2-10.

(3) Conduct training according to the following requirements and complete the work content in Table 2-2. When α=0° and α=60°, find the effective value of the load line voltage U_{UV} and draw its waveform, and draw the waveform of the U-phase line current at the same time.

Table 2-2 Comparison of current waveform between α=0° and α=60°

Control angle α	Effective value of the load line voltage U_{UV}	Line voltage waveform	Line current waveform
0°			
60°			

After completing the work content of Table 2-2, analyze the reason for the large difference between α=0° and α=60° waveforms.

3. Task procedures

Task procedures are as follows.

(1) Connect the isolation transformer in the Dy11 connection, and the synchronous transformer in the Yy10 connection with no load connected. Connect their primary side to a 220 V/127 V power supply. Use an oscilloscope to observe whether the amplitude and waveform of U_{A1}, U_A, and U_{SA} are normal. Use a multimeter to measure whether the +12 V, +15 V and +24 V power supply voltages are normal. The voltage U_C is supplied by an external stabilized power supply.

(2) Cut off the power supply, connect the upper end of the three-phase voltage regulating circuit to the isolation transformer to measure 47 V for the second time (low voltage is safer), and connect the lower end to a three-phase triangular (incandescent lamp) load.

(3) Trigger circuit unit, connect the power supply, connect the synchronization voltage, first

make U_C=4V, observe the pulse of N1 with an oscilloscope (whether it is a double pulsing with an interval of 60°), and then adjust U_C to the minimum (U_C=0), then adjust R_{P11} till the pulse of N1 disappears. (The amplitude of the sawtooth wave voltage is about 7.8−7.9 V).

(4) Adjust R_{P12} and R_{P13} till the slope of the sawtooth wave of N2 and N3 is the same as that of N1.

(5) Measure whether the output pulses of N1, N2 and N3, and the output pulses (in amplitude and waveform) by each pulse transformer are normal. If they are normal, disconnect the power supply and connect the output of the pulse transformer to the corresponding G and K poles of VT_1−VT_6.

(6) Turn on the power and adjust U_C under α=0° or α=60° respectively. Use a meter and an oscilloscope to measure the effective value of the line voltage, the line voltage waveforms, and line current waveforms.

4. Precautions

Precautions are as follows.

(1) Before training, it is necessary to preview the relevant knowledge part of this module, and carefully read the "Training Content" and "Task Procedures".

(2) Since there are many connections in the training circuit, one unit should be connected and checked to test whether it is normal. Only when it is ensured that the units are working properly, can the units be connected.

(3) There are many units in the training circuit that need to be compared and measured with an oscilloscope, and attention should be paid to find out the wiring of the common end of the two probes, otherwise it is easy to cause a short circuit.

(4) When use the two probes of the dual comprehensive oscilloscope to measure at the same time, the ground terminals of the two probes must be the terminals of the same potential (because the ground terminals of the two probes of the oscilloscope are connected together), otherwise, there will be a short circuit during the measurement.

(5) Since the common end of the oscilloscope probe is connected to the shell, and the shell is connected to the earth through the plug, and the neutral line of the three-phase power line is connected to the earth, then the ground wire of the probe is connected to the power neutral line. In the power electronics circuit training, if the probe is used to measure the thyristor components, it will burn out the components or cause a short circuit (if the rectifier transformer is not used). Therefore, it is usually necessary to remove the ground wire of the oscilloscope, or power the oscilloscope through an isolation transformer (for example, the sockets of the device should pass through an isolation transformer).

5. Notes and summary

____

____

____

Task 3　Comprehensive training of PWM controlled switching regulated power supply

✍ Relevant knowledge

1. PWM technology

PWM (Pulse Width Modulation) technology can be described as follows: modulating the width of a series of pulses to equivalently obtain the desired waveform (including shape and amplitude).

The application of PWM technology for inverter circuit is an important progress in the power electronics technology history. Various inverter circuits in use now apply PWM technology. Therefore, the training of combinating PWM technology and inverter circuit technology will enable us to have a complete understanding of the inverter circuit.

There is an important conclusion in the sampling control theory: when narrow pulses in equal impulse and different shapes are applied to the inertia part, their effects are basically the same. PWM technology is based on this conclusion to conduct (turn-on and turn-off) semiconductor switching devices, so that the output terminal gets a series of pulses with equal amplitude but unequal width, and these pulses are used to replace sine waves or other required waveforms. The width of each pulse is modulated according to certain rules, and the output voltage of the inverter circuit and the output frequency of that can be changed. Various narrow pulse models are shown in Figure 2-12.

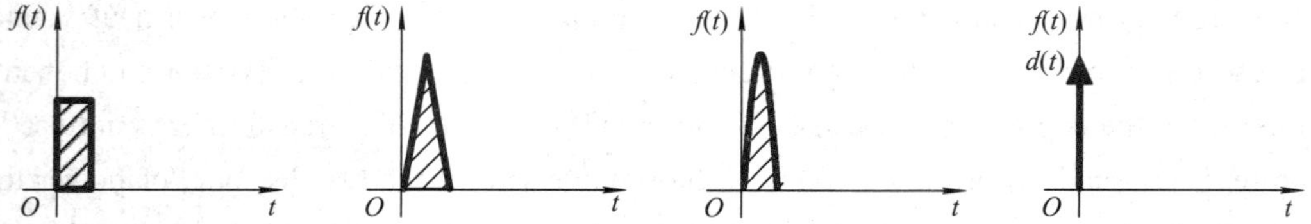

(a) Rectangular pulse　(b) Triangular pulse　(c) Sine half wave pulse　(d) Unit impulse function

Figure 2-12　Various narrow pulses with models

SPWM (Sinusoidal PWM) is a relatively mature and widely used PWM technology at present. The pulse width change according to the sine law to get the PWM waveform equivalent to the sine waveform (the SPWM waveform which can control the turn-on and turn-off of the switching devices in the inverter circuit), so that the area of the output pulse voltage is equal to that of the desired output sine wave in corresponding interval.

Divide the sine half wave pulse as shown in the Figure 2-12(c) into N equal parts, and then the sine half wave pulse can be regarded as a waveform composed of N pulse in sequences connecting to each other. These pulses have the same width and are equal to π/N, but the amplitudes are different, and the top of the pulse is not a horizontal straight line but a curve, and the amplitude of each pulse changes according to the sine law. If the above pulse in sequence is replaced with the same number of rectangular pulses with equal amplitude but unequal width, the midpoint of the rectangular pulse coincides with the midpoint of the corresponding sine wave, and the area (impulse) of the rectangular pulse and the corresponding sine wave is equal, then we get the PWM waveform. It can be seen that the amplitude of each pulse is equal, and the width changes according to the sine law. According to the principle of area equivalence, the sine half wave pulse and the PWM waveform are equal in value, as shown in Figure 2-13. For the negative half cycle of the sine wave, the same method can be used to obtain the PWM waveform. The PWM waveform whose pulse width changes according to the sine law and is equivalent to the sine wave is also called SPWM waveform.

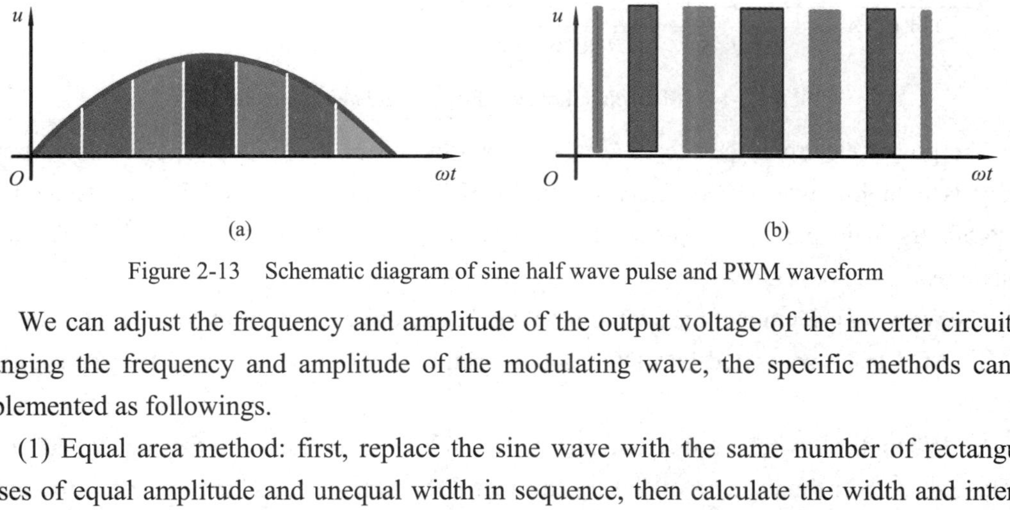

Figure 2-13 Schematic diagram of sine half wave pulse and PWM waveform

We can adjust the frequency and amplitude of the output voltage of the inverter circuit by changing the frequency and amplitude of the modulating wave, the specific methods can be implemented as followings.

(1) Equal area method: first, replace the sine wave with the same number of rectangular pulses of equal amplitude and unequal width in sequence, then calculate the width and interval of each pulse, and store these data in computer to generate the PWM signal (controlling the turn-on and turn-off of the switching device).

(2) Hardware modulation method: the principle of this method is to use the desired waveform as the modulating signal, and use the modulated signal as the carrier wave, modulating the carrier wave to obtain the desired PWM waveform. Usually an isosceles triangle wave is used as the carrier wave, when the modulating signal wave turns into the sine wave, we will get the SPWM waveform.

(3) Natural sampling method: it uses sine wave as modulating wave and isosceles triangle wave as carrier wave, compare the two waves and control the turn-on and turn-off of the switching device at the natural intersection of the two waveforms.

(4) Regular sampling method: the principle of this method is to use a triangle wave to sample a sine wave to obtain a ladder wave, and then use the intersection of the ladder wave and the triangle wave to control the turn-on and turn-off of the switching device, then we can realize the SPWM method. When the triangle wave is only at its top (or bottom) position, it starts to sample the sine wave, the pulse width determined by the intersection point of the step wave and the triangle wave is symmetrical within a carrier wave period (sampling period).

2. Switch-type regulated power supply

1) The main circuit of the switch-type regulated power supply

The main circuit of the switch-type (switching) regulated power supply is shown in Figure 2-14.

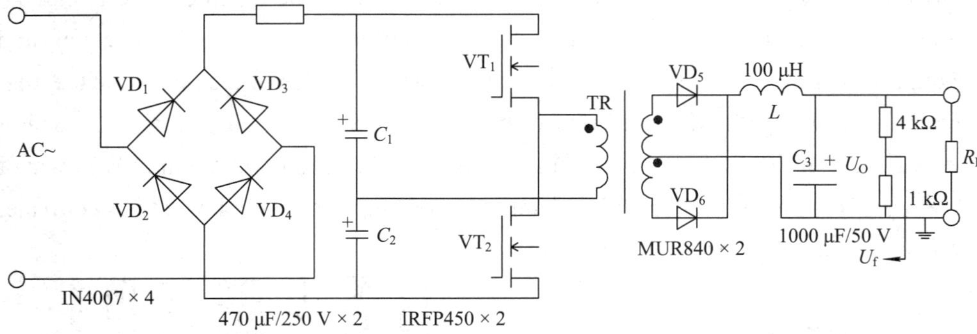

Figure 2-14　Main circuit of switch-type regulated power supply

The power electronic device used in the main circuit of the switch-type regulated power supply is a high-power MOSFET produced by IR Company in United States. Its model is IRFP450. Its main parameters are as follows: rated current 16 A, rated voltage 500 V, and on-state resistance 0.4 Ω.

The working principle of the switch-type regulated power supply is as follows.

① AC can be converted into DC through bridge rectification to charge the capacitors C_1 and C_2.

② When the switch tube (field effect tube) VT_1 is in turn-on, the voltage of C_1 is discharged to the primary side of the transformer TR (current flows in from the same ends); when VT_2 is turned on, the voltage of C_2 is also discharged to the primary side of TR (but the current is flow in from the opposite ends).

③ In this way, an alternating current is formed in the primary side of TR; this alternating current is transformed by a transformer, and then after full-wave rectification, and L and C filtering, it becomes a direct current, and its voltage is U_O.

④ By adjusting the duty cycle of the driving pulses of the switching tubes, VT_1 and VT_2 can adjust the output DC voltage.

U_f is a sampling voltage proportional to the output voltage, which is sent to the control terminal of the drive module as a negative voltage feedback signal.

2) Regulated power supply control circuit

The principle diagram of the regulated power supply control circuit formed by the field effect tube integrated drive circuit SG3525 is shown in Figure 2-15.

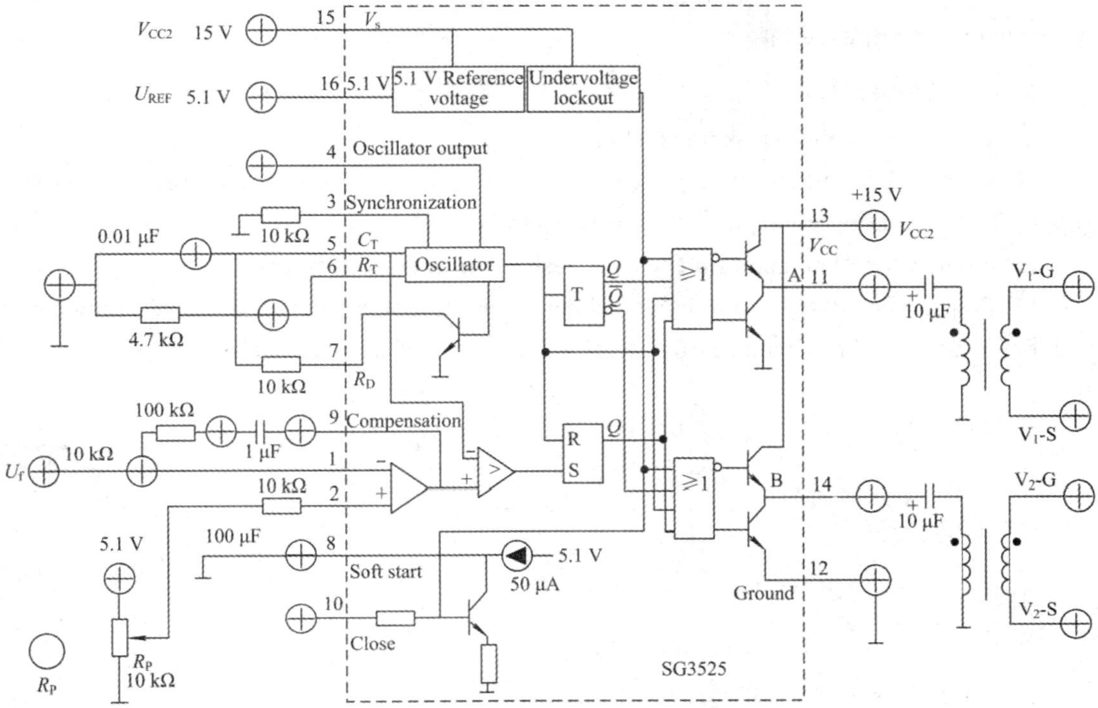

Figure 2-15 Regulated power supply control circuit

SG3525 is a dedicated PWM control circuit produced by Silicon General in United States. It adopts constant frequency pulse width modulation and is suitable for the control of various switch-type power supplies. SG3525 includes precision reference source, sawtooth oscillator, error amplifier, comparator, frequency divider, etc., it can realize the basic PWM control circuit, and contains a protection circuit.

(1) The frequency of SG3525 is mainly determined by R_T, C_T and R_D. The frequency f of its oscillator is

$$f=1/C_T(0.67R_T+1.3R_D)$$

Its input voltage V_{CC} is 8–35 V, usually 15V; output current is less than 40 mA.

(2) The functions of each pin of SG3525 are as follows: 1—Error Amplifier (AE) (inverting input); 2—AE (non-inverting input); 3—Synchronous voltage input; 4—Oscillator output; 5—External C_T; 6—External R_T; 7—External R_D; 8—Soft start; 9—Compensation; 10—Close (protection); 11—Output A; 12—Ground; 13—Connect to V_{CC2}; 14—Output B; 15—Connect to V_{CC2}; 16—Connect to the reference voltage U_{REF}.

(3) It can be seen from Figure 2-15 that the terminal 2 of the error amplifier (here as a voltage regulator) is connected to a given voltage, which must be given by 5.1 V via a 10 kΩ potentiometer (RP). Terminal 1 is connected to the negative feedback voltage U_f, and between the terminals 9 and 1, a resistor and a capacitor are connected in series. Therefore, the error amplifier can keep the voltage constant, thus forming a regulated power supply.

(4) Adjust RP, that is, adjust the duty cycle of the output square wave pulse, thus adjust the

output voltage of the regulated power supply.

✍ Practice and operation

1. Task preparations

There are three steps in task preparations.

(1) Clarify the working principle of the main circuit of the switch-type regulated power supply and the working principle of the PWM control circuit.

(2) Preview the “Training Content”and “Task Procedures” in this task.

(3) Prepare the “PWM controlled switching regulated power supply” part, wires, multimeter and oscilloscope. Refer to Figure 2-16 and Table 2-1 for related device models.

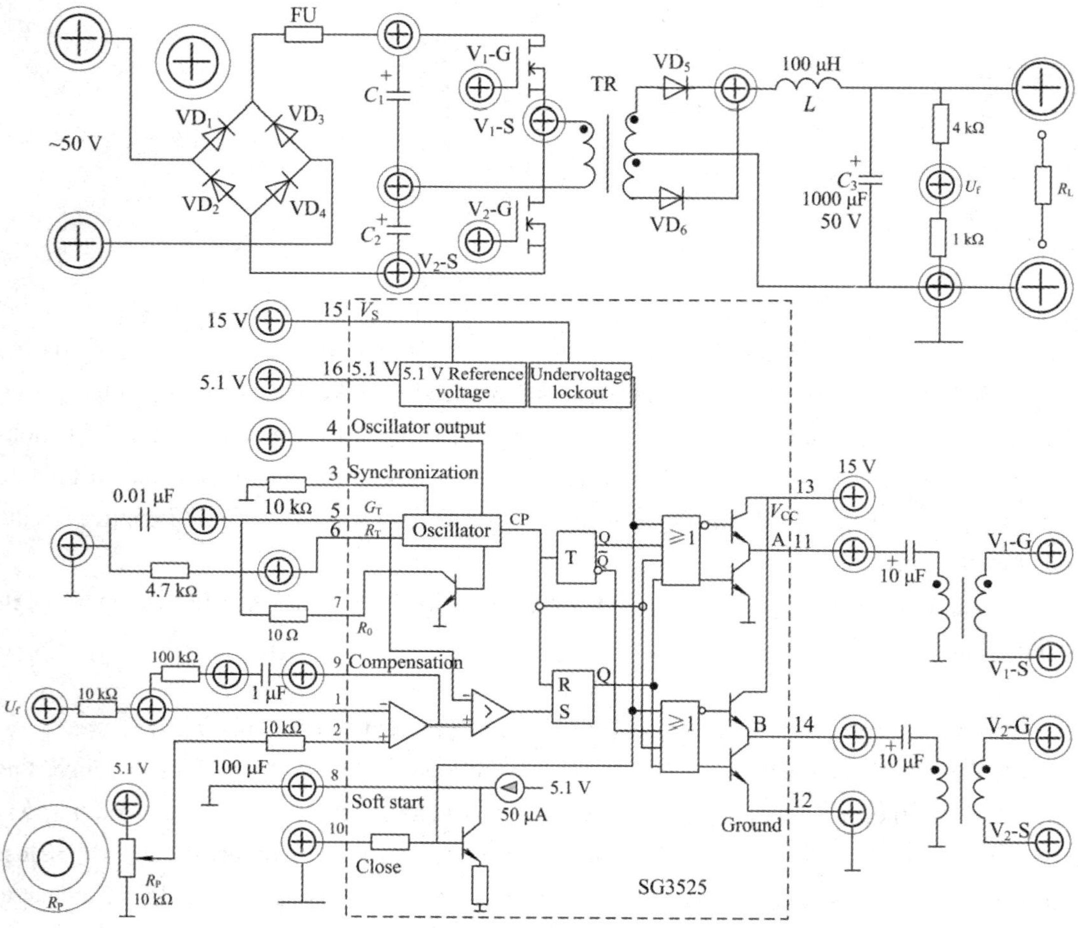

Figure 2-16 The training circuit of PWM controlled regulated power supply

2. Training content

The requirements of training content are as follows.

(1) Master the working principle of the training circuit in Figure 2-16, and prepare the circuits and related components required for the task.

(2) Complete the circuit construction according to Figure 2-16.

(3) Observe the waveform of the PWM signal, and record and analyze it.

① Will changing the RP change the pulse frequency, why?

② Why are the voltage waveforms different for different loads?

③ In Figure 2-16, if the 1μF capacitor between terminal 9 and terminal 1 is in short-circuited, what impact will it have on the circuit?

3. Task procedures

Task procedures are as follows:

(1) Turn on the power supply of the control circuit, use an oscilloscope to observe the sawtooth wave and the waveforms of the A and B PWM signals respectively, and record the waveform, frequency and amplitude.

(2) Turn on the main circuit power supply, observe the U_{GS} waveforms of the gate-source voltage and the drain-source voltage U_{DS} waveforms of the two MOSFETs, and record the waveform, period, pulse width and amplitude.

(3) Connect the resistance load (R_L=3 Ω and R_L=200 Ω), adjust the potentiometer R_P, so that the output voltage U_O is in 0, $U_{Om}/4$, $U_{Om}/2$, $2U_{Om}/3$ and U_{Om} five grades (U_{Om} is the maximum output DC voltage). Record the voltage waveform, period and amplitude under different loads and different output voltages.

4. Precautions

Precautions are as follows:

(1) Since the numerical values and waveforms in the power electronics circuit training are complicated and involve many factors, it is necessary to understand and master the working principle of the circuit and the numerical values and waveforms to be measured in the training. Avoid blindness in training.

(2) Other precautions are the same as Task 2 in this module.

5. Notes and summary

____

____

____

____

【Extended Practices】

Exercise 1 Comprehensive training of BJT single-phase parallel inverter circuit

1. Training preparations

There are two aspects of training preparation.

(1) Clarify the working principle of a single-phase parallel inverter circuit composed of power Bipolar Junction Transistors (BJT).

(2) Prepare the “single-phase parallel inverter circuit”, including connecting wire, multimeter, and oscilloscope. For related device models, see Figure 2-17 and Table 2-1.

2. Training content

Training content and requirements are as follows.

(1) Master the working principle of the training circuit in Figure 2-17 and prepare the module circuit and related components required for the task.

(2) Complete the circuit construction according to Figure 2-17.

(3) Observe the voltage waveforms at points 3, 4, 5, 6 and 7 of the control circuit, and record the load voltage, and the value and waveform of the input current of the inverter circuit.

The BJT single-phase parallel inverter circuit is shown in Figure 2-17.

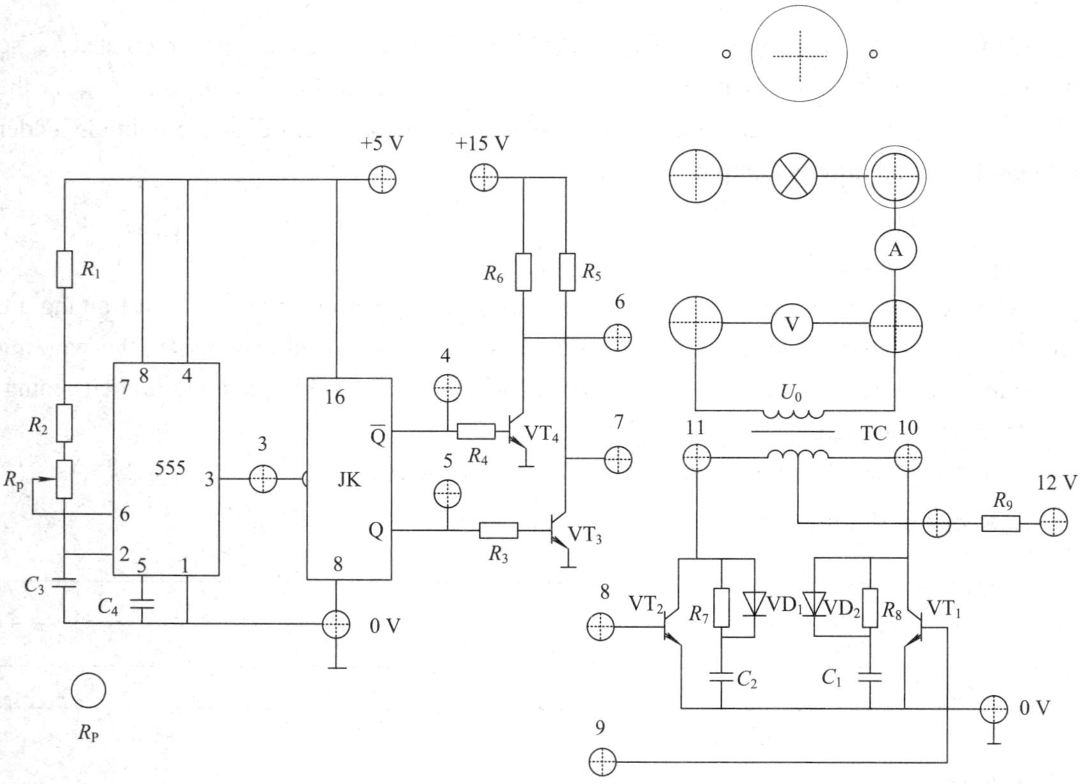

Figure 2-17　BJT single-phase parallel inverter circuit

1) The working principle of the training circuit

The training circuit consists of two parts: the pulse generating circuit (control circuit) and the inverter circuit (main circuit). Its detailed composition is introduced as follows.

① The circuit formed by the 555 timer is a multivibrator. Adjust the potentiometer R_P, the frequency of the output will change accordingly. Similarly, when the frequency changes in this circuit, the duty cycle will also change (and the duty cycle q>50%).

② The JK flip-flop in Figure 2-17 is a shaping circuit that protects VT_3 and VT_4. In VT_3 and VT_4, only one of them can be in a conducting state (prevent inverter failure). The amplifier circuit is composed of VT_1 (and VT_2).

③ A single-phase (passive) inverter circuit is formed by power transistors VT_1, VT_2 and transformer TC. The resistance-capacitance and fast recovery diodes connected in parallel with VT_1 and VT_2 form the energy-dissipating turn-off buffer (absorption) circuit to alleviate the impact when the transistor is suddenly turn-off. R_9 in the circuit is a protection resistor to prevent excessive current from forming when the inverter fails (after the circuit is working in normal state, short-circuit R_9).

3. Task procedures

Task procedures are as follows:

(1) Connect the +15 V and +5 V power supplies to the control circuit, and use an oscilloscope to observe the value and waveform of the voltage at each point (3, 4, 5, 6, 7) of the control circuit. The main points of observation are as follows:

① Adjust R_P. Is the frequency continuously adjustable? What is the frequency at this time? Does the pulse width change when the frequency changes?

② Whether the frequency at points 4 and 5 is half the frequency of point 3, and whether the waveforms at points 4 and 5 are exactly opposite?

③ Whether the waveform and amplitude at points 6 and 7 are the same as those at points 4 and 5?

(2) Connect the +12V power supply in the main circuit (due to the large current, it is recommended to use a DC adjustable power supply), voltmeter, ammeter and load (incandescent lamp), and connect the main circuit with the control circuit.

(3) Measure the voltage waveform on the load with an oscilloscope and observe whether the inverter circuit is working normally.

The observation essentials are as follows:

① Observe voltage waveform between 11 and 12 points (or 10 and 12 points).

② Observe voltmeter and ammeter readings.

③ Observe the voltage waveform on the load (incandescent lamp).

If the inverter circuit is working normally, short-circuit R_9.

(1) Adjust R_P, and record the value and waveform of load voltage U_0 and inverter circuit input current I when R_P is zero (frequency $f=f_0$) and R_P is maximum ($f=f_m$) in Table 2-3.

Table 2-3 Value and waveform record table

Observations		U_0/V	I/A	U_0 wave	I wave
Frequency/Hz	f_0				
	f_m				

4. Precautions

Precautions are as follows:

(1) When measuring and comparing the waveform of each point with a dual-trace oscilloscope, the common ends of the probes are all connected to the circuit zero (G) end.

(2) The current-limiting resistor R_9 can be removed (short-circuited) after the circuit is working normally.

5. Notes and summary

____

____

____

____

____

Exercise 2　Comprehensive training of given integral circuit

1. Training preparations

(1) Familiar with the working principle of a given integral circuit (also known as a given integrator).

(2) Prepare the "given integral circuit", such as connecting wire, multimeter, and oscilloscope. For related device models, see Figure 2-18 and Table 2-1.

2. Training content

Training content and requirement are as follows.

(1) Master the working principle of the training circuit in Figure 2-18, and prepare the module circuits and related components required for the task.

(2) Complete the circuit construction according to Figure 2-18.

(3) Observe and record the values and waveforms of the output signal.

The given integral circuit is shown in Figure 2-18. Its circuit analysis includes the following aspects:

(1) The given integral circuit is composed of a voltage comparator (operational amplifier) N_1, an integrator N_2, and an inverter N_3. The comparator input loop resistance R_1=30 kΩ, and the feedback resistance R_2=1 MΩ, so N_1 has a very high gain. Therefore, a very small input voltage can make N_1 reach the limit value, so that N_1 can be the comparator. The clamp diodes VD_1 and VD_2 above and below N_1 form a negative feedback limiter circuit. Adjust the potentiometers R_{P3} and R_{P4} to set the negative and positive limit values.

(2) The limiting voltage U_{1m} output by the voltage comparator N_1 is sent to the integrator N_2, and the integral regulator output voltage u_2 (that is, U_{02}) is inverted by the inverter N_3 to become U_{01}, and U_{01} is fed back to the input terminal of N_1. Since the polarity of U_{01} is exactly opposite to the polarity of the given integral input u_{i1}, negative feedback is formed.

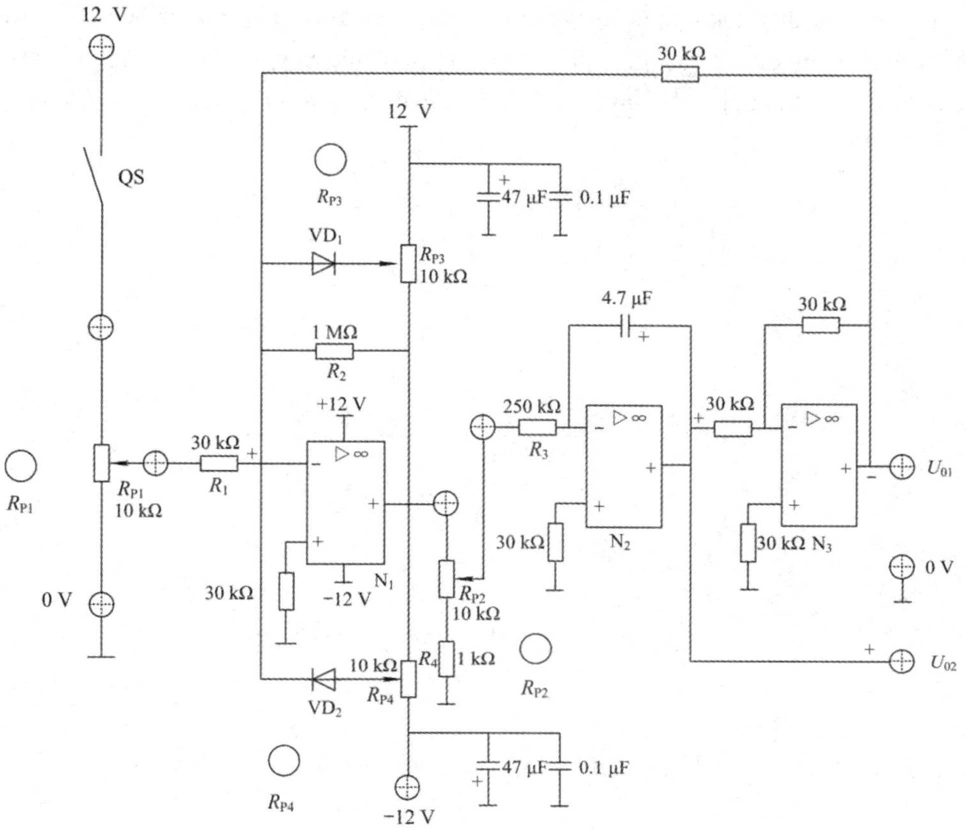

Figure 2-18 Given integral circuit

(3) The output-input curve of a given integrator is shown in Figure 2-19. When the given integral input U_i is a step signal and large enough, such as a few tenths of volts or more. Then the output of N$_1$ quickly reaches the negative limit value U_{1m}, and this voltage is divided by the resistor R_4 and the potentiometer R_{P2} (u_1) into the integrator (see Figure 2-19(b)). The output voltage of the integrator $u_2 = \frac{1}{R_5 C}\int u_1 \mathrm{d}t$, since u_1 is a constant quantity, $u_2 = \frac{1}{R_5 C}\int u_1 \mathrm{d}t = \frac{-u_1}{R_5 C}t$. It can be seen from this formula that it is an oblique straight line with a slope $m = -\frac{1}{R_5 C}$. Now u_i is a positive value, and the comparator N$_1$ is input from the inverting terminal, so u_1 is a negative value, so from the previous formula, we can see that m is a positive value and u_2 is a rising oblique straight line (see the initial section of Figure 2-19). Adjust R_{P4}, you can adjust u_1 and the rising slope.

When u_2 rises to the input voltage U_{i1}, it is fed back to the N$_1$ input terminal after the inverter N$_3$, and the current input from the outside at the N$_1$ terminal at this time $i_1 = \frac{U_{i1}}{R_1} + \frac{U_{01}}{R_1} = 0$ (because at this time u_i=U_{i1}, U_{01}=$-u_2$=$-U_{i1}$). Then the output u_1 of N$_1$

becomes zero, and then the output u_2 of the integrator remains to be the value of U_{i1}. (See the middle section of Figure 2-19(c)). If the output is u_2 of the integrator, it is U_{02} (as shown in Figure 2-19(c)). If it is output by inverter, it is U_{01}, and its polarity is opposite to U_{02} (as shown in Figure 2-19(d)).

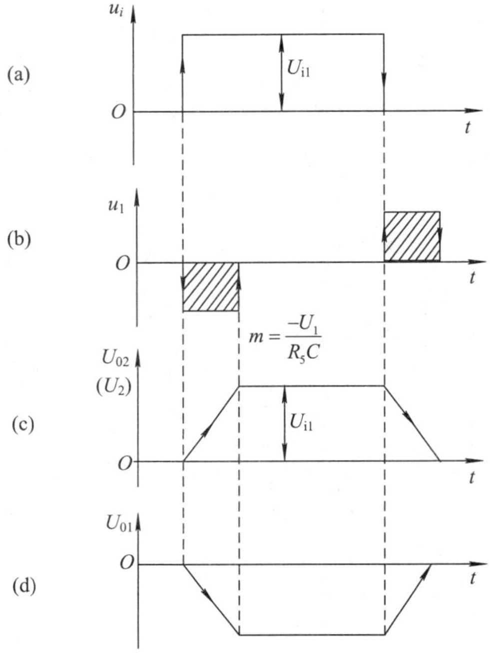

Figure 2-19　The output-input curve of a given integrator

When the machine stops, in the same way, when u_i is adjusted to the initial value (0V), the voltage at the input of N_1 is only u_1 (its polarity is opposite to U_{i1}), which will make N_1 grow to the positive limit value quickly. This voltage is divided after the voltage is pressed, it is input to the integrator, so that u_2 gradually decreases until u_2=0, and the voltage at the input terminal of N_1 returns to zero again. Adjust R_{P3}, you can adjust the reduced power.

It can be seen from the above analysis that the step signal can be converted into a ramp signal through a given integrator. Adjust the potentiometers R_{P4} and R_{P3} to set the rising (or falling) slope of the output signal, that is, the acceleration of the speed regulation system (speed rise rate or fall rate) is set. Figure 2-20 shows the speed starting curve of a given integrator.

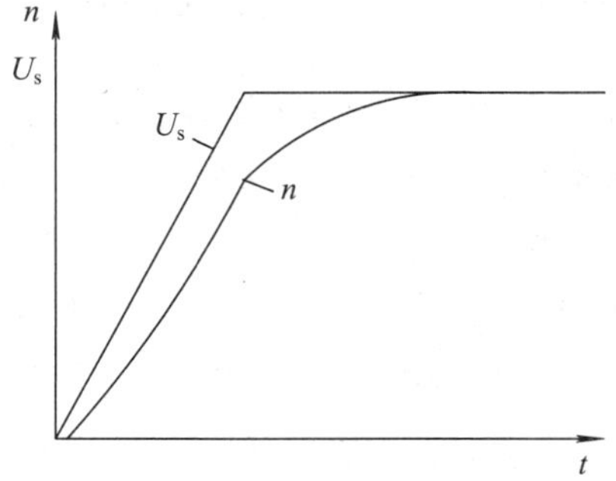

Figure 2-20　Speed starting curve of a given integrator

3. Task procedures

Task procedures are as follows:

(1) The given integral circuit is connected to the ±12V power supply and the common neutral line.

(2) Suddenly close the switch QS to form a step signal. Use a slow sweep oscilloscope to observe the waveforms of u_1, u_2 (i.e. U_{02}).

(3) Adjust the potentiometer R_{P2} and observe its influence on the slope of the ramp.

(4) Adjust the potentiometer R_{P3} and observe its influence on the output voltage curve.

(5) Adjust the potential R_{P1} and observe its influence on the output voltage curve.

4. Precautions

Precautions are as follows:

(1) Use a slow-scan oscilloscope to measure dynamic curves. If you use a general dual trace oscilloscope, you can only see the measured amplitude (horizontal line) moving slowly up (or down). When use a dual trace oscilloscope, pay attention to when measuring different voltages for comparison, first find the common terminal (0 V terminal) to facilitate the connection of the common terminals of the two probes.

(2) When the voltage at the input of the steady-state integral regulator is zero, the output will remain at the original value.

5. Notes and summary

____

____

____

____

____

____

Module 3 Comprehensive training of electric drive system

【General Introduction】

Electric drive (also known as motor drive) refers to a type of drive that uses a motor as the prime mover to drive the production of mechanical motion. It can realize the conversion between electrical energy and mechanical energy, and conveniently control the motor output shaft, the torque, angular acceleration, speed, angular displacement and various starting, running, speed changing, braking, etc. of the dragged machinery or mechanical combined load according to the requirements of the production process. This module mainly involves the AC electric drive system. The frequency converter is used to realize the speed regulation operation of the electric drive system.

This module mainly selects Siemens MM420(frequency inverter). The MM420 (MICRO MASTER 420) inverter series adopts high-performance voltage and frequency control technology to provide low-speed high-torque output and good dynamic characteristics. At the same time, it has a strong overload capacity to meet a wide range of requirements under application occasions. For the application of the inverter, you must first be familiar with the operation panel of the inverter, and then set various function parameters of the inverter according to the actual application.

This module is divided into three tasks:

(1) Basic training of MM420 inverter fast speed regulation.

(2) Basic training of MM420 inverter multi-stage speed regulation.

(3) Training of MM420 inverter analog smooth speed regulation.

【Training Objectives】

1. Knowledge objectives

(1) Master the circuit structure of the MM420 inverter.

(2) Master the composition and working principle of MM420 inverter.

(3) Master the parameter setting methods of MM420 inverter.

2. Skill objectives

(1) Able to analyze and select reasonable inverter parameters for different training tasks.

(2) Able to set correct parameters for the inverter.

(3) Able to wire the inverter terminals correctly.

(4) Able to debug the inverter according to the requirements of the training tasks.

3. Operating standards

1) Preparation before training

(1) Carefully read the Siemens MM420 inverter user manual, and make notes on related content. Do a good knowledge preview in advance, and thoroughly understand the relevant theoretical knowledge.

(2) Clarify the training objectives of knowledge and skill, and go to the training room in advance to understand the structural characteristics of the training apparatus, accessories and instruments required for the training tasks.

(3) Clarify and remember the training safety standards to avoid personal safety accidents.

In summary, doing a good job of preparatory work before training is a prerequisite for cultivating students' good habits to complete experimental tasks.

2) Task procedures

(1) Spare parts. Prepare the devices required for the training according to the device preparation list for each task.

(2) Recognize pictures. Correctly understand the electrical control schematic diagram given in the training task, recognize the electrical symbols of all components in the diagram, and clearly correspond to the actual components in the drawing.

(3) Wiring. Strictly follow the wiring diagram of the inverter in the training task.

(4) Inspection and investigation. Do an electrical test on the well-connected training circuit, focus on checking whether there is a short circuit or open circuit.

(5) Operation. Read the inverter user manual carefully, learn more about the parameters that need to be set, and master the correct setting methods of the inverter parameters. After the parameter setting is completed, perform the operation of the inverter in sequence.

(6) Disassembly of the training task and reset of the device. After the training is completed, the state of the training equipment should be restored in time. The main circuit and control circuit power supply must be turn-off before disconnecting, and then wait a little time before disconnecting, and remove the wires, instruments and components. Organize them and place them in the specified place.

3) Training report

Writing a training report is a process for students to summarize and improve after completing the training task. In the training report, the operation process needs to be recorded, and the working principle of the training task should be explained in detail, and the experience of completing the training task should be written.

4. Safety standards

1) Inspection and preparation before operation

The following should be carefully checked before the inverter starts running.

(1) Check whether the wiring is correct, especially check the output terminal UVW of the

inverter, this terminal cannot be connected to the main power supply, and confirm that the grounding terminal is well grounded.

(2) Confirm that there is no short-circuit or short-circuit to ground in the equipment or exposed live parts between the terminals.

(3) Confirm that the terminal connection wires (connectors) and screws are all tight.

(4) Confirm that the motor is not connected to other mechanical loads.

(5) Before connect to the power supply, make all the switches in the off position to ensure that the inverter cannot start, and no abnormal actions occur.

2) Maintenance and inspection

In order to ensure the long-term reliable and continuous operation of the inverter, regular inspections should be performed.

(1) Daily inspection. Do not remove the outer cover of the inverter when the inverter is power-on for operation. Check the operation state from outside to confirm that there is no abnormality. The following aspects are usually checked.

① The operating performance meets the standard specifications.

② The surrounding environment meets the standard requirements and is in normal.

③ The display panel is normal.

④ There is no abnormal noise, vibration and peculiar smell.

⑤ There is no abnormality such as overheating or discoloration.

(2) Regular inspections. The regular inspection items and methods are shown in the Table 3-1 as follows.

Table 3-1　Periodic inspection items and methods

Check part	Check item	Method	Standard
Surroundings	Confirm the ambient temperature, vibration	Measure with sight and instrument	Compliance with technical specifications
Display panel of keyboard	Whether the displayed content is legible and whether there are missing characters	Measure with sight	Display panel reading is without abnormalities
Frequency converter	No abnormal vibration and odor	Hearing, sight, smell	No abnormality
Cooling system	No abnormal sound, looseness, peculiar smell and discoloration	Measure with sight	No abnormality in the balance rotation

5. Operation quality testing

The quality standards of training are as follows.

(1) During the training process, the inverter circuit should be connected in strict accordance with the wiring diagram.

(2) Set correct parameters for inverter in each task.

(3) During the training process, check whether the operations conform to the quality requirements and operating standards under rigorous attitude.

【Training Requirements】

1. Preparations of tools, accessories and instruments

Tools, accessories and instruments are shown in Table 3-2 below.

Table 3-2 Tools, accessories and instruments

No.	Name	Unit	Amount
1	MM420 frequency converter	Set	1
2	Motor	Set	1
3	Air switch	Piece	1
4	Button	Piece	3
5	Wire		Appropriate amount
6	Multi-turn sliding potentiometer	Piece	1
7	Electrician's toolkit	Set	1

2. Knowledge assessment point

(1) The composition and circuit structure of the MM420 inverter.

(2) The working principle of MM420 frequency converter.

(3) The general parameter meaning of MM420 inverter.

(4) Parameter selection for multi-stage speed regulation with MM420 frequency converter.

(5) Parameter selection for analog smooth speed regulation with MM420 frequency converter.

3. Skills assessment point

(1) The panel operation of the MM420 inverter.

(2) The digital terminal wiring of the MM420 inverter.

(3) The analog terminal wiring of the MM420 inverter.

(4) Parameter setting and operation for multi-speed operation with MM420 frequency inverter.

(5) Parameter setting and operation for analog smooth speed regulation with MM420 frequency inverter.

【Work Tasks】

Task 1 Basic training of fast speed regulation of MM420 inverter

✍ Relevant knowledge

Frequency conversion and speed regulation is a method to change the frequency of the motor's stator power supply, thereby to change its synchronous speed. The main equipment of the frequency conversion and speed regulation system is the frequency converter providing the frequency conversion power. The frequency converter can be divided into two types: the AC-DC-AC inverter and the AC-AC inverter. At present, mostly used frequency inverter in China is the AC-DC-AC inverter.

(1) The frequency converter can be divided into four parts. The general frequency converter is composed of the main circuit and the control loop. The power conversion part providing the voltage modulation and frequency modulation power supply to the asynchronous motor is called the main circuit. The main circuit includes a rectifier, a smoothing circuit (also called an intermediate DC link) and an inverter.

① Rectifier. Its function is to transform the industrial frequency power supply into a DC power supply;

② Smoothing circuit. Since the load of the inverter is an asynchronous motor, it is an inductive load. Regardless of whether the motor is in electric state or in generating state, the initial power factor will never be equal to 1. Therefore, there will always be an exchange of reactive power between the intermediate DC link and the motor, and this kind of reactive energy depends on the energy storage components(capacitors or inductors) of the intermediate DC link for buffering, so the intermediate DC link is actually an intermediate energy storage link.

③ Inverter. In contrast to the role of the rectifier, the inverter converts DC power into AC power at the required frequency. The structure of the inverter is a three-phase bridge inverter circuit composed of 6 semiconductor switching devices. By regularly controlling the turn-on and turn-off of the main switching control of the inverter, a three-phase AC output waveform at any frequency can be obtained.

(2) Control loop is often composed of arithmetic circuits, detection circuits, control signal input and output circuits, drive circuits and brake circuits. Its main task is to complete the switching control of the inverter, the voltage control of the rectifier, and the completion of various protection functions.

Common models of Siemens general-purpose inverters are MM440, MM430, MM420, G120C, etc. The frequency inverter equipment used in this training is the Siemens MM420 inverter, which is controlled by a microprocessor and uses Insulated Gate Bipolar Transistors (IGBT) with modern advanced technology as the power output device. These transistors have high operational reliability and function diversity. The switching frequency of pulse width modulation is optional, which helps to control the noise of motor operation.

1. Inverter external terminal

The three-phase power supply of the MM420 converter is connected to L1, L2 and L3. The output terminals are U, V and W. The circuit must be grounding. The main circuit wiring of MM420 inverter is shown in Figure 3-1.

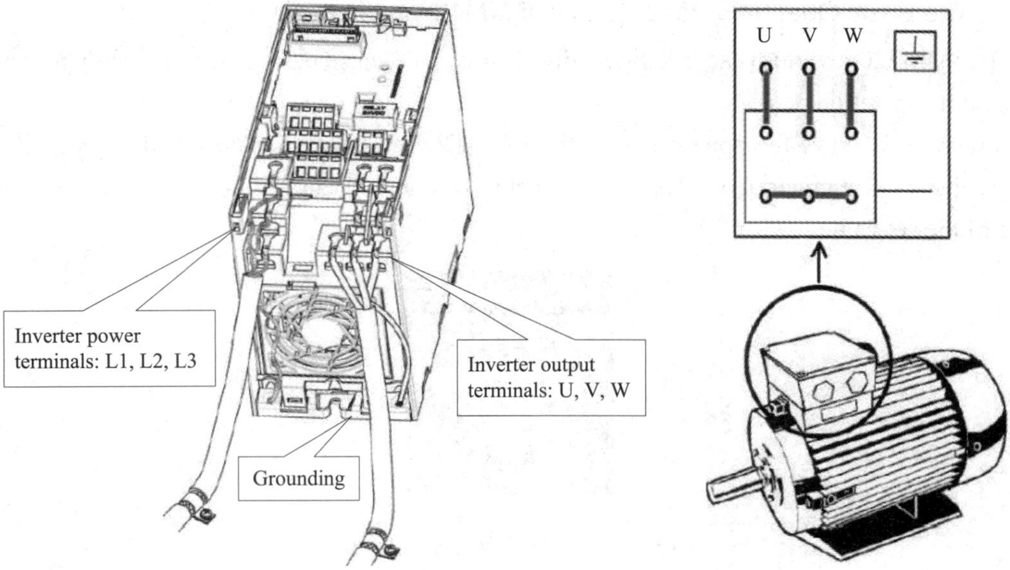

图 3-1 Main circuit wiring of MM420 inverter

The wiring terminals and functions of the MM420 inverter are shown in Figure 3-2. It has a human-computer interaction interface—Basic Operation Panel (BOP). Its core component is the CPU unit. According to the set parameters, the controllable sine wave signal of MM420 inverter is output by calculation, and it is modulated by SPWM(Sinusoidal Pulse Width Modulation). The amplified output three-phase AC voltage drives the three-phase AC motor to run.

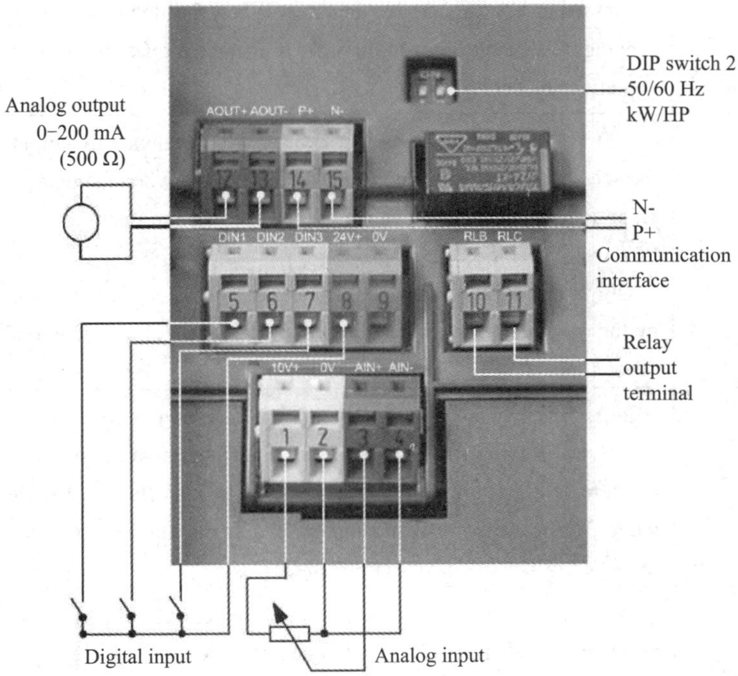

1—10 V DC voltage output
2—0 V (the grounding terminal of 10 V DC voltage output)
3—Positive voltage terminal of analog input
4—Negative voltage terminal of analog input
5, 6, 7—Digital quantity (switch quantity) input terminal
8—+24 V DC voltage output
9—0 V (that is, the grounding voltage of 24 V DC voltage)
10, 11— Relay output terminal
12, 13— Analog output terminal
14, 15— RS485 serial communication interface; +10 V DC voltage output

Figure 3-2 Wiring terminals and functions of MM420 inverter

2. The Basic Operating Panel (BOP) of MM420 inverter

The MM420 inverter is an intelligent digital inverter, whose basic parameter settings can be made on its operation panel.

Figure 3-3 shows the appearance of the BOP. BOP has five display positions, which can display the serial number and values of parameters, alarm and fault information, and the actual value of the set value.

Figure 3-3 The appearance of the BOP

The buttons on the BOP and their functions are shown in Table 3-3.

Table 3-3 The buttons on the BOP and their functions

Display/Button	Function	Function description
	Status display	LCD displays the current set value of the inverter
	Start key	Press this key to start the inverter. When the default value is running, this key is blocked. In order to make the operation of this key effective, set P0700=1
	Stop key	OFF1: Press this key, and the inverter will decelerate to stop at the selected ramp down rate. This key is blocked when the default value is running; in order to allow this key to be in operation, set P0700=1. OFF2: Press this key twice (or once, but for a longer time), the motor will freely stop under the action of inertia. This function is always "enabled"
	Change the direction of rotation of the motor	Press this key to change the rotation direction of the motor. When the motor is reversed, the key is indicated by a minus sign, or there is a flashing decimal point used to indicate the default value. When the motor is in running, this key is blocked. In order to make this key effective, set P0700=1
	Motor jogging	When the inverter has no output, pressing this key will start the motor to run at the preset jog frequency. Releasing this key will make the inverter stop. If the inverter/motor is in running, pressing this key will have no effect

Continued table

Display/Button	Function	Function description
	Function key	1. Browse auxiliary information function(press for 2 seconds). When the inverter is running, press this key when displaying any parameter and keep it for 2 seconds, the following parameter values will be displayed (the parameter values can start to display randomly when the inverter is running): (1) DC link voltage (represented by d, its unit is V). (2) Output current (A). (3) Output frequency (Hz). (4) Output voltage (represented by o, its unit is V). (5) The value selected by P0005 (if P0005 selects to display any of the above parameters, it will no longer be displayed here). Pressing this key several times will display the above parameters in turn. 2. Jump function(press for a short time) When any parameter (rXXXX or PXXXX) is displayed, press this key for a short time, and it will jump to r0000 immediately. If necessary, you can continue to modify other parameters. After jumping to r0000, press this key to return to the original display point
	Access key	Press this key to have access to the parameter setting
	Increase key	Press this key to increase the parameter value displayed on the panel
	Decrease key	Press this key to decrease the parameter value displayed on the panel

3. MM420 inverter parameter setting example (taking P0004 modification as an example)

BOP can be used to modify and set the system parameters, so that the inverter has the desired characteristics, the steps to change the value of any parameter can be roughly summarized as follows.

(1) Find the selected parameter number.

(2) Enter the parameter value access level and modify the parameter value.

(3) Confirm and store the modified parameter value.

The specific steps for modifying P0004 parameter value are shown in Table 3-4.

Table 3-4 The specific steps for modifying P0004 parameter value

Steps	Display value
(1) Press""to access parameter setting	r0000
(2) Press "" until P0004 is displayed	P0004
(3) Press "" to enter the parameter value access level	0
(4) Press ""or ""to reach the required values	3
(5) Press "" to confirm and store the values of the parameters	P0004
(6) Users can only see the command parameters	

✍ Practice and operation

1. Task preparations

Prepare components used in this training task and fill in Table 3-5 as required.

Table 3-5 Components list

No.	Name	Unit	Amount
1	MM420 converter	Set	1
2	Motor	Set	1
3	Air switch	Piece	1
4	Button	Piece	3
5	Wire		Appropriate amount
6	Multi-turn sliding potentiometer	Piece	1
7	DC power supply	Piece	1
8	Electrician's Toolkit	Set	1

2. Training content

1) Inverter wiring

The wiring schematic diagram of the main circuit of the inverter is shown in Figure 3-4, and the physical wiring map of the inverter main circuit is shown in Figure 3-5.

Practitioners must check the inverter manual before starting wiring, confirm whether the inverter matches the power supply voltage. It is not allowed to connect single-phase and three-phase 220 V inverters to 380 V or higher power supplies.

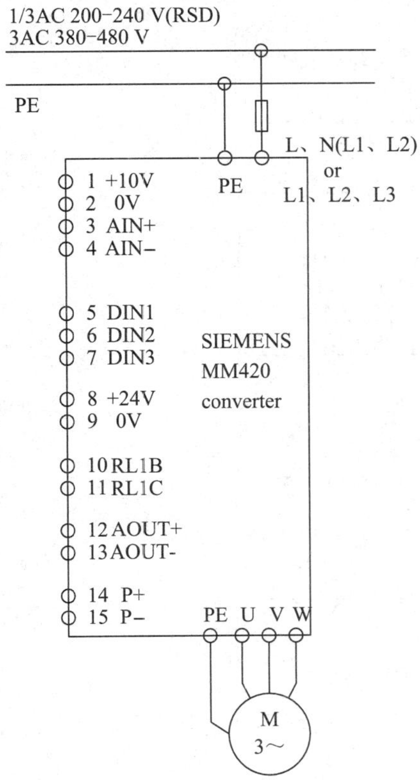

Figure 3-4 Wiring schematic diagram of the main circuit of the inverter

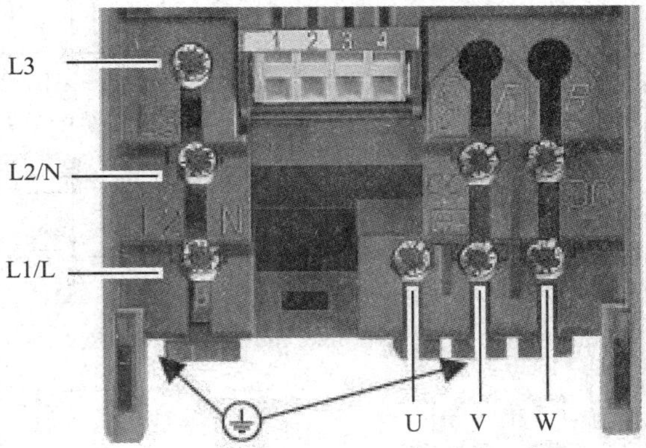

Figure 3-5 Physical wiring map of inverter main circuit

2) Reset the inverter to the factory default setting

In order to reset all the parameters of the inverter to the factory default settings, the parameters should be set according to the following values:

① Set P0010=30;

② Set P0970=1.

It should be noted that after setting the keys, wait for about 1 minute for the completion of the reset process.

3) Quick debugging of the inverter

Use the quick debugging function to match the inverter with the actually-used motor parameters, and set important technical parameters. If the rated data of the motor stored in the frequency converter is consistent with the data on the nameplate configuration of the motor commonly-used, quick debugging is unnecessary. The flow chart of quick debugging is shown in Figure 3-6.

P0010: Start quick debugging
0: Ready to run
1: Quick debugging
30: Factory default settings instruction
Before the motor is put into operation, P0010 must be set as '0'.However, if P3900=1 is selected after debugging, then the operation of "P0010 equals 0" automatically

↓

P0100: Choose work area is Europe/North America
0: The power unit is kW: the default value of f is 50 Hz
1: The power unit is HP: the default value of f is 60 Hz
2: The power unit is kW: the default value of f is 60 Hz

↓

P0304: Rated motor voltage is 10–2000 V
(Rated motor voltage is entered according to the nameplate configuration of the motor)

↓

P0305: Rated motor current
(Nominal motor current in A from rating plate)

↓

P0307: Rated motor power
(Nominal motor power in kW/HP from nameplate)
If P0100 = 0 or 2, value will be in kW. If P0100 = 1, value will be in in HP(1HP=0.735 kW)

↓

P0310: Rated motor frequency
(Nominal motor frequency in Hz from nameplate)
Pole pair number is recalculated automatically if parameter is changed

↓

P0311: Rated motor speed
(Nominal motor speed in r/min from nameplate)
Setting 0 causes internal calculation of value

↓

P0700: Selection of command source
0: Factory default setting
1: BOP (keypad)
2: Terminal input

P1000: Selection of frequency setpoint
1: MOP setpoint 2: Analog setpoint
3: Fixed frequency 4: USS on BOP link
5: USS on COM link

↓

P1080: Minimum frequency (Enter the minimum motor frequency 0–650 Hz)
Set minimum motor frequency at which motor will run at a speed being irrespective of frequency setpoint. The value set here is valid for both clockwise and anticlockwise rotation of the motor

↓

P1082: Max. frequency(enters the maximum motor frequency 0–650 Hz)
Set maximum motor frequency at which motor will run at a speed being irrespective of the frequency setpoint. The value set here is valid for both clockwise and anticlockwise rotation of the motor

↓

P1120: Ramp-up time(enters the ramp-up time in s)
Time taken for motor to accelerate from standstill up to maximum motor frequency (P1082) when no gounding is used

↓

P1121: Ramp-down time(enters the deceleration time in s)
Time taken for motor to decelerate from maximum motor frequency (P1082) down to standstill when no gounding is used

↓

P3900: End of quick debugging
0: End of quick debugging, no motor calculations or reset to factory default setting
1: End of quick debugging, motor calculations and reset to factory default setting
2: End of quick motor calculations and I/O reset
3: End of quick debugging, motor calculations but no I/O reset

Figure 3-6 Quick debugging flow chart

3. Task procedures

Task procedures are as follows:

(1) Inverter startup. Press the key  on the front operation panel of the inverter, and the inverter will start to drive the motor to increase speed and run at the speed corresponding to the frequency set by P1040.

(2) Forward/reverse rotation and acceleration/deceleration operation. The motor speed (running frequency) and direction of rotation can be directly changed by pressing the increase

key/decrease key (▲/▼) on the front operation panel.

(3) Jog operation. Press the jog key "jog" on the front operation panel of the inverter, and the inverter will drive the motor to increase speed and run at the positive jog frequency value set by P1058. When the jog key "jog" on the operation panel of the frequency inverter is released, the frequency converter will drive the motor speed down to zero. Then, if you press the reversing key on the operation panel of the inverter and repeat the above-mentioned jog operation, the motor can be jog to run in the reverse direction under the drive of the inverter.

(4) Motor stop. Press the stop key "0" on the front operation panel of the frequency inverter, and the frequency inverter will drive the motor speed down to zero.

4. Precautions

After operating the panel keys to make the inverter run, check the following items:

(1) Whether the rotation direction of the motor meets the requirements.

(2) Whether the rotation direction of the motor is stable.

After above test operation, it is confirmed that there is no abnormal situation, and the inverter can be put into operation immediately.

5. Notes and summary

____

____

____

____

____

____

Task 2 Basic training of multi-stage speed regulation of MM420 inverter

✍ Relevant knowledge

The multi-stage speed regulation, also called fixed frequency, is to use the switch terminal to select the combination of fixed frequency under the condition of parameter setting P1000=3 to realize the multi-stage speed operation of the motor. Multi-stage speed regulation can be achieved by the following three methods.

1. Terminal direct selection

Set "P0701−P0703 = 15", that is, select the "terminal direct selection" mode in multi-speed regulation. In this mode, a digital input selects a fixed frequency. The terminal and parameter setting corresponds to Table 3-6.

Table 3-6 Terminal and parameter setting

Terminal number	Corresponding parameters	Corresponding frequency setting value	Explanation
5	P0701	P1001	(1) The frequency reference source P1000 must be set to 3 (2) When multiple frequency selections are activated at the same time, the selected frequency is the sum of them
6	P0702	P1002	
7	P0703	P1003	

2. Terminal direct selection + ON command

Set "P0701–P0703 = 16", that is, select the "terminal direct selection + ON command" mode. In this mode, the digital input signal has both the frequency selection function and the motor start function.

3. Binary code selection + ON command

Set "P0701–P0703 = 17", that is, the "binary code + ON command" mode is selected. In this mode, the digital input signal composes the binary code to select the fixed frequency. The frequency of each frequency band is set by the parameters P1001–P1007, and up to 7 frequency bands can be controlled. The numerical selection of each fixed frequency is shown in Table 3-7. In multi-band control, the direction of the motor speed is determined by the positive-negative frequency set by the parameters P1001–P1007. There are 3 digital input ports, which one is used as motor running and stopping control, and which is used as multi-frequency control, can be determined by the user arbitrarily. Once the control function of a certain digital input port is determined, its internal parameter setting value must be consistent with the port corresponding to the control function. The fixed frequency selection is shown in Table 3-7.

Table 3-7 Fixed frequency selection

Frequency setting	DIN3	DIN2	DIN1
P1001	0	0	1
P1002	0	1	0
P1003	0	1	1
P1004	1	0	0
P1005	1	0	1
P1006	1	1	0
P1007	1	1	1

✍ Practice and operation

1. Task preparations

Prepare a list of components used in this training task as Table 3-8.

Table 3-8 Component list used in the task

No.	Name	Unit	Amount
1	MM420 inverter	Set	1
2	Motor	Set	1
3	Air switch	Piece	1
4	Button	Piece	3
5	Wire		Appropriate amount
6	Multi-turn sliding potentiometer	Piece	1
7	DC power supply	Piece	1
8	Electrician's toolkit	Set	1

2. Training content

1) Inverter wiring

Connect the inverter according to the inverter wiring diagram in Figure 3-7.

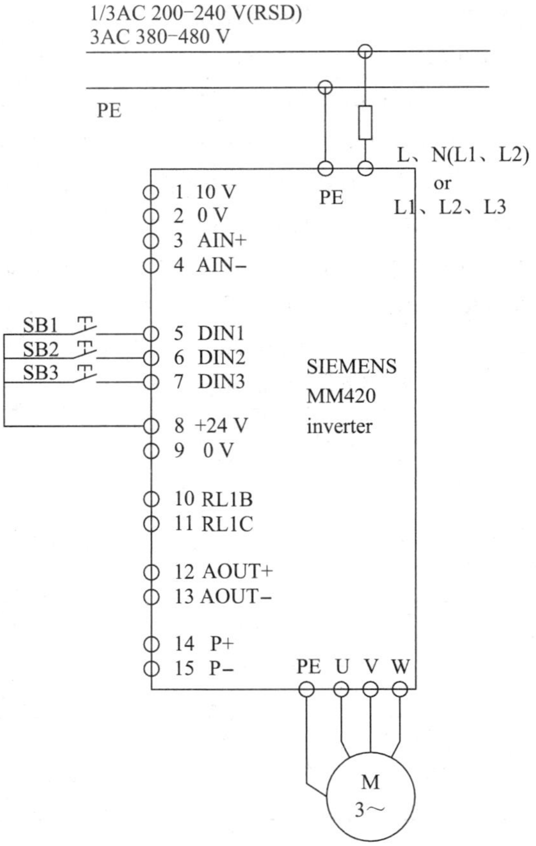

Figure 3-7 Schematic diagram of inverter wiring

2) Inverter parameter setting

The parameters of the inverter are set as follows:

(1) Correctly set the rated frequency, rated voltage, rated current, rated power, and rated

speed of the inverter output.

(2) Achieve the motor's multi-stage speed regulation through the external terminals, the on and off state combination of switches "SB1" "SB2" and "SB3" are set differently through using the operation panel for running frequency regulation and acceleration/deceleration time change of the motor.

3. Task procedures

Task procedures are as follows:

(1) Check whether the equipment in the training equipment is complete.

(2) Complete the wiring of the inverter according to the external wiring diagram of the inverter, and check carefully to ensure that it is correct.

(3) Turn on the power switch and set the inverter parameters correctly according to the parameter setting Table 3-9.

Table 3-9 Inverter parameter setting

No.	Inverter parameters	Set value	Function Description
1	P0304	According to the nameplate configuration of the motor	Rated voltage of the motor (V)
2	P0305	According to the nameplate configuration of the motor	Motor rated current (A)
3	P0307	According to the nameplate configuration of the motor	Motor rated power (kW)
4	P0310	According to the nameplate configuration of the motor	Motor rated frequency (Hz)
5	P0311	According to the nameplate configuration of the motor	Motor rated speed (r/min)
6	P1000	3	Fixed frequency setting
7	P1080	10	The minimum frequency of the motor(0 Hz)
8	P1082	50.00	Maximum frequency of the motor(50 Hz)
9	P1120	3	Ramp-up time(10 s)
10	P1121	3	Ramp-down time(10 s)
11	P0700	2	Select the command source (controlled by the terminal block)
12	P0701	17	Set value of fixed frequency (binary code selection + ON command)
13	P0702	17	Set value of fixed frequency (binary code selection + ON command)
14	P0703	17	Set value of fixed frequency (binary code selection + ON command)
15	P1001	5.00	Fixed frequency 1
16	P1002	10.00	Fixed frequency 2

Continued table

No.	Inverter parameters	Set value	Function Description
17	P1003	20.00	Fixed frequency 3
18	P1004	25.00	Fixed frequency 4
19	P1005	30.00	Fixed frequency 5
20	P1006	40.00	Fixed frequency 6
21	P1007	50.00	Fixed frequency 7

(4) Change the “on” and “off” state of switches “SB1”, “SB2”, “SB3”, observe and record the output frequency of the inverter shown in Table 3-10.

Table 3-10 The output frequency of the inverter

SB1	SB2	SB3	Output frequency
OFF	OFF	OFF	
ON	OFF	OFF	
OFF	ON	OFF	
ON	ON	OFF	
OFF	OFF	ON	
ON	OFF	ON	
OFF	ON	ON	
ON	ON	ON	

4. Precautions

Precautions are as follows:

(1) Before set the values of parameters, reset the values of inverter parameters to the factory default settings, set P0010=30 and P0970=1, press the P key to start the reset.

(2) Set P0003=2 to have access to extended parameters.

(3) When set the motor parameters, first set P0010=1 (quick debugging), and then set P0010=0 (preparation) after complete the motor parameter setting.

5. Notes and summary

____

____

____

____

____

Task 3 Training of analog smooth speed regulation of MM420 inverter

✍ Relevant knowledge

MM420 frequency inverter (aka MM420 inverter) can control the motor forward and reverse running, and forward and reverse jog running direction through 3 digital input ports. You can increase and decrease the output frequency through the basic operation panel and press the frequency adjustment button to set the forward and reverse speed. The speed of the motor can also be controlled by the analog input. The purpose of this task is to control the speed of the motor through the analog value of the analog input terminal.

The output terminals "1" and "2" of the MM420 inverter provide a high-precision 10 V DC power supply for the user's given unit. The speed adjustment potentiometer can be used in series in the circuit to adjust the potentiometer and change the analog input voltage given by the input port AIN1+. The input value of the frequency inverter will closely follow the change of the given value, so as to smoothly adjust the value of the motor speed.

MM420 inverter provides users with analog input ports, namely ports "3" and "4". By setting the parameter value of P0701, the digital input port "5" has the function of forward rotation control; by setting the parameter value of P0702, the digital input port "6" has a reverse control function; the analog input ports "3" and "4" are connected with an external potentiometer, and an adjustable analog voltage signal is input through the port "3" to control the value of the motor speed. That is, the direction of the motor speed is controlled by the digital input, and the value of the speed is controlled by the analog input.

✍ Practice and operation

1. Task preparations

Prepare the list of components used in this training task and fill in the Table 3-11 as required.

Table 3-11 Components list used in the task

No.	Name	Unit	Amount
1	MM420 frequency inverter	Set	1
2	Motor	Set	1
3	Air switch	Piece	1
4	Button	Piece	3
5	Wire		Appropriate amount
6	Multi-turn sliding potentiometer	Piece	1
7	DC power supply	Piece	1
8	Electrician's toolkit	Set	1

2. Training content

1) Inverter wiring

Connect the inverter according to Figure 3-8.

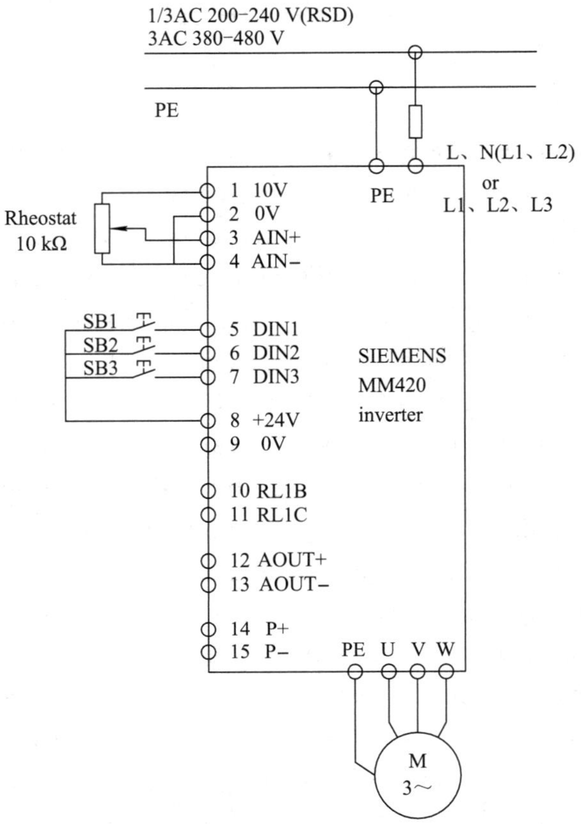

Figure 3-8 Wiring diagram of the inverter

2) Parameter setting

The relevant parameters of the inverter are set as follows:

(1) Restore the factory default settings of the inverter, set P0010=30 and P0970=1, and press the P key to start resetting.

(2) Set the motor parameters as shown in Table 3-12. After setting the motor parameters, set P0010=0, and the inverter is currently in the ready state, and it can run normally.

Table 3-12 Motor parameter settings

Parameter number	Factory default value	Setting value	Description
P0003	1	1	Set user access level as standard level
P0010	0	1	Quick debugging
P0100	0	0	Working area: The power is expressed in kW, and the frequency is in 50 Hz
P0304	230	230	Motor rated voltage(V)

Continued table

Parameter number	Factory default value	Setting value	Description
P0305	3.25	0.9	Motor rated current(A)
P0307	0.75	0.4	Motor rated power(kW)
P0308	0	0.8	Motor rated power(cosϕ)
P0310	50	50	Motor rated frequency (Hz)
P0311	0	1400	Motor rated speed (r/min)

(3) Set the analog signal operation control parameters as shown in Table 3-13.

Table 3-13 Analog signal operation control parameters setting

Parameter number	Factory default value	Setting value	Description
P0003	1	1	Set user access level as standard level
P0004	0	7	Command and digital I/O
P0700	2	2	Command source selection is input by terminal block
P0003	1	2	Set user access level as extended level
P0004	0	7	Command and digital I/O
P0701	1	1	ON refers to the motor is running forward, OFF refers to stop running
P0702	1	2	ON refers to the motor is running reverse, OFF refers to stop running
P0003	1	1	Set user access level as standard level
P0004	0	10	Set point channel and ramp function generator
P1000	2	2	Frequency setting value is selected as analog input
P1080	0	0	Minimum frequency of motor operation (Hz)
P1082	50	50	The highest frequency of motor operation (Hz)

3) Inverter running operation

(1) Motor forward rotation and speed regulation. Press the motor forward self-locking button SB1. The digital input port DIN1 is “ON”. The motor is running forward. The speed is controlled by the external potentiometer RP1. The analog voltage signal changes in the range of 0−10 V, correspondingly the frequency of the inverter changes in the range of 0−50 Hz, corresponding to the motor speed change is in the range of 0−1500 r/min. When the lock button SB1 is released, the motor stops running.

(2) Motor reverse rotation and speed regulation. Press the motor reverse self-locking button SB2, and the digital input port DIN2 is “ON”, the motor runs in reverse, which is the same as the motor forward, and the reverse speed is still adjusted by the external potentiometer. When the self-locking button SB2 is released, the motor stops running.

3. Task procedures

Task procedures are shown as follows:

(1) Check whether the equipment in the training equipment is complete.

(2) Complete the wiring of the inverter according to the external wiring diagram of the inverter, and check carefully to ensure that the wiring is correct.

(3) Turn on the power switch and set the inverter parameters correctly according to the parameter setting in Table 3-13.

(4) Press the button "SB1" to start the inverter in forward direction, and adjust the frequency rotation potentiometer to change the output frequency of the inverter.

(5) Press the button "SB3" to stop the inverter.

(6) Press the button "SB2" to start the inverter in reverse direction, and adjust the frequency rotation potentiometer to change the output frequency of the inverter.

(7) Press the button "SB3" to stop the inverter.

4. Precautions

Precautions are shown as follows:

(1) Before set the parameters, reset the inverter parameters to the factory default settings, set P0010=30 and P0970=1, and press the P key to start resetting.

(2) Set P0003=2 to allow access to extended parameters.

(3) When set the motor parameters, first set P0010=1 (in quick debugging mode); after the motor parameter setting is complete, set P0010=0 (in preparation mode).

5. Notes and summary

____

____

____

____

【Extended Practices】

Exercise 1 Motor start-stop operation training based on PLC and inverter

1. Training preparations

Prepare components according to the following requirements:

(1) Tools: safety connection line.

(2) Meter: digital multimeter.

(3) PLC: One unit of Siemens S7-200 host.

(4) One Siemens programming cable.
(5) One MM420 inverter.
(6) One three-phase asynchronous motor.
(7) One computer (PC).

2. Training content

Control the start and stop of the motor through the button input terminal of PLC. Set a forward button and a reverse button to connect to the PLC input terminal, and the PLC outputs a digital signal to the external digital terminal control end of the MM420 inverter to achieve the purpose of controlling the operation of the motor .

1) Analyze control requirements and assign input address/output address

The PLC input address /output address allocation is shown in Table 3-14.

Table 3-14 Input Address /Output Address Allocation

Input	I0.0
Description	Start up button
Output	Q0.0
Description	Start up the inverter

2) PLC control principle diagram

According to Table 3-14, follow the I/O wiring diagram of the PLC controlled inverter as shown in Figure 3-9 for the wiring of the training task.

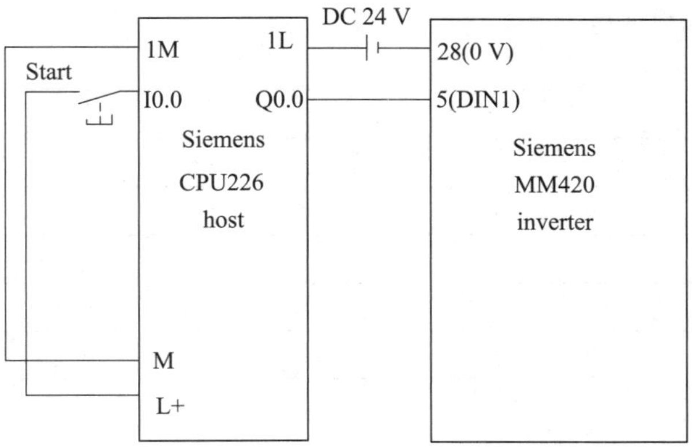

Figure 3-9 I/O wiring diagram of PLC controlled inverter

3) Setting of inverter parameters

First, restore the factory settings of the inverter by referring to the setting method of factory reset parameter P0970. Then enter the quick debugging (P0010=1) to check whether the motor parameters in the frequency inverter are related to the currently-used motor, if not, we need to correct it. (motor rated voltage P0304=230 V, motor rated current P0305=0.9 A, motor rated power P0307=0.4kW, motor rated frequency P0310=50Hz, motor rated speed P0311=1400 r/min, command source selection=0.9A, P0700=2, frequency setting P1000=1). Finally, end the quick debugging (P3900=1). Then set parameter P0003=2 (access level) and P0701=1(forward rotation).

3. Task procedures

(1) Check whether the components in the training is complete.

(2) Wiring according to the external wiring diagram of the PLC controlled inverter, carefully check to ensure that wiring is correct.

(3) Turn on the power switch and set the inverter parameters correctly according to the parameter setting in Table 3-13.

(4) Download the PLC control program to the unit of the host.

(5) Press the start button to start the inverter and observe whether the motor rotates normally.

4. Precautions

Precautions are as follows:

(1) Before set the parameters, reset the inverter parameters to the factory default settings, and set P0010=30 and P0970=1, and press the P key to start resetting.

(2) Set P0003=2 to have access to extended parameters.

(3) When setting the motor parameters, first set P0010=1 (in quick debugging made), and then set P0010=0 (in preparation mode) after completing the motor parameter setting.

(4) When downloading the PLC program, ensure that the download line is connected correctly and the communication can be successful.

5. Notes and summary

____

____

____

____

____

____

____

Exercise 2　Based on PLC, inverter motor forward and reverse running training

1. Training preparations

Prepare components according to the following requirements:

(1) Tools: safety connection line.

(2) Meter: digital multimeter.

(3) PLC: One unit of Siemens S7-200 host.

(4) One Siemens programming cable.

(5) One MM420 inverter.

(6) One three-phase asynchronous motor.

(7) One computer (PC).

2. Training content

Control the forward and reverse running (rotation) of the motor through the button input to the PLC. Set a forward button and a reverse button to connect to the PLC input terminal. The PLC outputs 2 digital signals to the external digital control terminal of the MM420 inverter. In this way, the forward and reverse rotation of the motor can be controlled.

1) Analyze control requirements and assign input address/output address

The PLC input address/output address allocation is shown in Table 3-15.

Table 3-15 Input Address/Output Address Allocation of PLC

Input	I0.0	I0.1
Description	Inverter forward button	Inverter reverse button
Output	Q0.0	Q0.1
Description	Forward	Reverse

2) PLC external wiring

According to Table 3-15, follow the I/O wiring diagram of the PLC controlled inverter shown in Figure 3-10 for the wiring of the training task.

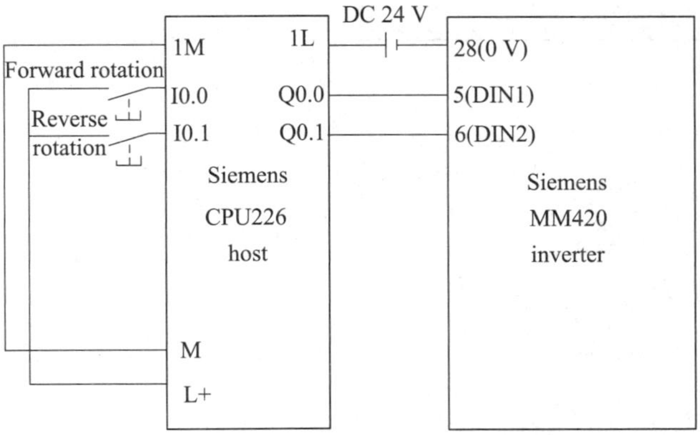

Figure 3-10 I/O wiring diagram of PLC controlled inverter

3) Setting of inverter parameters

First, restore the factory default settings of the inverter by referring the setting parameter P0970. Then enter the quick debugging (P0010=1) to check whether the motor parameters in the frequency inverter are related to the current motor, if not, we need to correct it. (motor rated voltage P0304=230 V. Motor rated current P0305=0.9 A, motor rated power P0307=0.4 kW, motor rated frequency P0310=50Hz, motor rated speed P0311=1400 r/min, command source selection P0700=2, frequency setting P1000=1). Finally, end the quick debugging (P3900=1). Then set the parameters P0003=2 (access level). P0701=1 (forward rotation) and P10702=2 (reverse rotation).

3. Task procedures

Task procedures are as follows:

(1) Check whether the equipment in the training equipment is complete.

(2) Wiring according to the external wiring diagram of the PLC controlled inverter, carefully check to ensure that wiring is correct.

(3) Turn on the power switch and set the relevant parameters of the inverter correctly according to the parameter setting in Table 3-12.

(4) Download the PLC control program to the unit of the host.

(5) Press the button SB1 to start the inverter and observe the direction of motor rotation.

(6) Press the button SB2 and observe the direction of motor rotation.

4. Precautions

Refer to Task 1 in this module.

5. Notes and summary

____

____

____

____

____

____

Module 4 Comprehensive training of programmable logic control system

【General Introduction】

S7-200 series PLC (referred as S7-200 PLC or S7-200 PLC control system) is a kind of small PLC system newly launched by SIEMENS company. With its compact structure, good scalability, powerful command function and low price, it has become an ideal controller for various contemporary small-scale control projects.

S7-200 PLC is composed of a host (basic unit), I/O expansion unit, functional unit (module) and external equipment. S7-200 PLC has two generations of products, CPU21X and CPU22X, among which CPU22X PLC series has four basic models: CPU221, CPU222, CPU224 and CPU226. This project selects CPU226 as the control unit, and through the design of four different tasks, hope to cultivate senior electricians' PLC application ability.

This module is mainly designed for training senior electricians. There are three tasks in total. The tasks are as follows:

(1) Programming and debugging of automatic feeding and loading control system.

(2) Programming and debugging of traffic light automatic control system.

(3) Programming and debugging of four-floor elevator control system.

Through the training of this module, students can achieve the ability to use S7-200 PLC to install, wire, program, debug and run the automation control system, and meet the skill requirements of senior electricians.

【Training Objectives】

1. Knowledge objectives

(1) Understand the working principle of S7-200 PLC control system.

(2) Master the distribution of digital I/O addresses of the S7-200 PLC control system.

(3) Master the usage of S7-200 PLC basic programming instructions.

(4) Master the programming method of S7-200 PLC control system.

2. Skill objectives

(1) Able to wire the digital input terminals of the S7-200 PLC control system.

(2) Able to wire the digital output terminals of the S7-200 PLC control system.

(3) Able to complete the programming and monitoring of the S7-200 PLC control program.

(4) Able to complete the commissioning and operation of tasks of S7-200 PLC control system.

3. Operating standards

1) Preparations before training

Before the training, you should do the following preparations.

(1) Carefully read the operation instruction book about the equipment, and at the same time review the relevant content in the textbook and make notes. Do a good knowledge preview in advance, and clarify the relevant theoretical knowledge.

(2) Clarify the training knowledge goals and ability goals, and go to the training room in advance to understand the structural characteristics of the training equipment and the list of components required for the training tasks.

(3) Clarify and remember the training precautions to avoid personal safety accidents and damage on equipment, non-standard operations must be prohibited.

In summary, doing a good job of preparatory work before training is a prerequisite for cultivating students' good habits and completing training tasks.

2) Task procedures

Task procedures are as follows:

(1) Spare parts. Prepare training equipment and consumables required for training according to the list.

(2) Drawing. Correctly understand the control requirements of the training task, and draw the PLC control electrical schematic diagram.

(3) Wiring. Observe the electrical wiring process and standards. Perform the installation layout and wiring of the components in accordance with the electrical schematic diagram of the training task.

(4) Inspection and investigation. Perform electrical testing on the connected training circuit, focus on checking whether there is a short circuit or open circuit.

(5) Programming, running and debugging. Power on the system, perform PLC programming and debugging, download the control program to the PLC, operate according to the operating principle, and complete the task of system operation.

(6) Disassembly of the training task and reset of the device. After the training is completed, it is often the time when accidents are most likely to occur due to a relaxed mood of the operator. Therefore, the main circuit and control circuit power supply must be turned off, and wait a little time before disconnecting the wires. Instruments and training devices should be arranged and placed in the specified place.

3) Training report

Writing a training report is a process for students to summarize and improve after completing the training task. In the training report, the operation process needs to be recorded, and the working principle of the training task should be explained in detail, and the experience of completing the training task should be written.

4. Safety standards

1) Short circuit protection

When the PLC output device is short-circuited, in order to avoid damage to the internal output components of the PLC, a fuse should be installed in the external output loop of the PLC for short-circuit protection. It is best to install a fuse in the circuit of each load.

2) Interlocking measures

In addition to ensuring the interlocking relationship of the circuits in the program, hardware interlocking measures should also be taken in the external wiring of the PLC to ensure the safe and reliable operation of the system. For example, the motor forward and reverse control should use the interlocked points of contactors KM1 and KM2 (normally-closed contacts) outside the PLC. When there are interlocking requirements between different motors, it is best to perform hardware interlocking measures outside the PLC. Interlocking with hardware outside the PLC is a common practice in PLC control systems.

3) Voltage loss protection and emergency shutdown measures

The power supply line of the PLC external load should have voltage loss protection measures(wiring method: when the power supply is restored after a temporary power failure, the external load of the PLC cannot start by itself without pressing the "Start" or "RUN" button). Another function of this wiring method is that when an emergency shutdown is required under special circumstances, the load power can be cut off by pressing the "STOP" button, which do no influence on PLC.

5. Quality standards

Quality standards are as follows:

(1) The training should be carried out in strict accordance with the task requirements, and the wiring should be carried out in accordance with the PLC control system schematic diagram given by the task to ensure that the PLC I/O terminal wiring is accurate.

(2) When in compiling a PLC control program, marks should be made in the program to improve the legibility of the program.

(3) During the training process, the teacher should test if the students' working attitude is positive, and if they follow quality requirements and operating standards.

【Training Requirements】

1. Preparations of tools, instruments and equipment

The tools, instruments and equipment that need to be prepared during the training are shown in the following Table 4-1.

Table 4-1　List of components

No.	Name	Unit	Amount
1	YL 360 PLC training platform	Set	1
2	Siemens S7-200 PLC host unit	Piece	1
3	Automatic feeding and loading control system for training	Piece	1
4	Traffic light automatic control system for training	Piece	1
5	Four-floor elevator control system for training	Piece	1
6	Siemens S7-200 programming cable	Piece	1
7	Computer	Set	1
8	Connection line	Piece	20

2. Knowledge identification point

(1) The application knowledge of Siemens S7-200 PLC basic instructions.
(2) The use of Siemens S7-200 PLC programming software.
(3) The programming method of Siemens S7-200 PLC.
(4) The debugging method of Siemens S7-200 PLC program.

3. Skills assessment point

(1) The I/O distribution of S7-200 PLC.
(2) The wiring and installation of S7-200 PLC.
(3) The programming for the tasks listed in this project.
(4) The debugging and operation of the programs for the tasks listed in this module.

【Work Tasks】

Task 1　Programming and debugging of automatic feeding and loading control system

✍ Relevant knowledge

The S7-200 PLC contains a single S7-200 CPU and various optional expansion blocks (units), which can be used to form different controllers easily in different size. The scale of the control points can range from several points to several hundred points. S7-200 PLC can easily form a PLC-PLC network and a microcomputer-PLC network to complete larger projects.

The S7-200 programming software STEP7-Micro/WIN SMART can easily program, debug, and monitor PLC in Windows environment, making PLC programming more convenient and faster. S7-200 can perfectly meet the requirements of various small-scale control systems.

The S7-200 CPU continuously executes user tasks in cyclic sequence is called scanning. It is divided into five stages: read input, execute program, process communication request, execute CPU self-diagnosis, and write output.

According to the main tasks, the scan cycle can be simplified into three stages: read input, execute user program, and write output.

S7-200CPU has two working modes:

(1) STOP working mode. When the CPU is in STOP working mode, the program is not executed. At this time, it is available to load program to the CPU or perform system settings.

(2) RUN working mode. The CPU runs the user program in the RUN working mode.

1. The composition of S7-200 PLC system

The S7-200 PLC system consists of a host, I/O expansion unit, text/graphic display, external devices, etc. Outside the CPU224 host box, there are an RS-485 communication interface connecting to external devices (such as programmers, text/graphic displays, PLC networks, etc), a working mode switch, analog potentiometer, I/O expansion interface, user program memory card, I/O wiring terminal block and working status indicator, etc.

1) Basic I/O

CPU22X series PLC has two different power voltages: 24 V DC and 85–264V AC. The output circuit is divided into relay output circuit and transistor DC output circuit. CPU22X series PLC can provide 10 basic unit CPUs (divided into 4 different types) for users to choose. The types and parameters are shown in Table 4-2.

Table 4-2 Types and parameters of CPU22X series PLC

	Type	Power supply	Input voltage	Output voltage	Output current
CPU221	DC input, DC output	24 V DC	24 V DC	24 V DC	0.75A, transistor
CPU222	DC input, Relay output	85–264 V AC	24 V DC	24 V DC, 24–230 V AC	2A, relay
CPU224	DC input, DC output	24 V DC	24 V DC	24 V DC	0.75A, transistor
CPU226/ CPU226XM	DC input, Relay output	85–264 V AC	24 V DC	24 V DC, 24V–230 V AC	2A, relay

2) I/O and expansion blocks of host

The I/O and expansion blocks of the CPU22X series PLC host are shown in the following Table 4-3.

Table 4-3　I/O and expansion blocks of the CPU22X series PLC host

Type	Host input points	Host output points	Scalable module
CPU221	6	4	None
CPU222	8	6	2
CPU224	14	10	7
CPU226	24	16	7

3) High speed reactivity

CPU224 PLC has 6 high-speed counting pulse input terminals (I0.0–I0.5), the fastest response speed is 30 kHz, used to capture pulse signals faster than the CPU scan cycle.

CPU224 PLC has two high-speed pulse output terminals (Q0.0–Q0.1), and the output pulse frequency can be as large as 20 kHz used for PTO and PWM high-speed pulse output.

4) Storage System

The S7-200 CPU memory system is composed of RAM and EEPROM, which are used to store user programs, CPU configuration, program data, etc.

5) Analog potentiometer

The analog potentiometer is used to change the value in the special register so as to change the program running parameters, such as the timing, the preset value of the counter, the processing quantity corresponding to control parameters, etc.

6) Memory card

The memory card can be optionally installed with expansion cards, which include blocks of EEPROM memory card, battery and clock card. The EEPROM memory card black is used for copying and duplicating user programs. The battery block is used to store data for a long time. The data storage time of the CPU224 internal storage capacitor is 190 hours, and the data storage time of the battery block can reach 200 days.

2. The internal components of S7-200 PLC

1) Data storage type and addressing method

The data storage type in the storage unit of the S7-200 series are Boolean, integer, and real types. The addressing method in S7-200 CPU is divided into three types: immediate addressing, direct addressing and indirect addressing.

(1) Immediate addressing. In an instruction, if the operand after the opcode is what specifically required by the opcode, and we call such instruction immediate addressing.

For example, the transfer instruction of Operand “MOV IN OUT” indicates that it transfers the data from in IN to in OUT, where IN is source operand, OUT is target operand.

If the transfer instruction is “MOVD 2505 VD500”. Its function can be described as follows. Transmit the decimal number 2505 to VD500, where 2505 is the source operand. Because the value of this operand is already in the instruction, there is no need to address it anymore, this operand is an immediate value. The value of the target operand is not given in the instruction, only the address VD500 to be transmitted is given. The addressing mode of this operand is direct addressing.

(2) Direct addressing. In an instruction, if the operand after the opcode appears in the form of address where the operand is located, this addressing method is called direct addressing.

For example: MOVD VD400 VD500.

Function: Transmit the double word data from in VD400 to in VD500.

(3) Indirect addressing. In an instruction, if the operand after the opcode appears in the form of address where the operand's address is located in it, the addressing method of this instruction is called indirect addressing.

For example: MOVD 2505 *VD500.

*VD500 refers to the address where the 2505 address is stored.

For example, VD0 is stored in VD500, and VD0 is the address of 2505.

The function of this instruction: Transmit the decimal number 2505 to the VD0 address.

2) Data storage area and component functions

The data storage area and component functions of S7-200 PLC are as follows.

(1) Input/output mapping register: S7-200 PLC addressing range (I0.0–I15.7) input mapping register (this area can be operated by bit, it can also be called as input relay). The input relay coil is driven by an external signal, normally-open-contact and normally-closed-contact are used for programming.

Output mapping register: S7-200 PLC addressing range (Q0.0–Q15.7). The output mapping register (also called output relay) is used to transmit the output signal of the PLC to the load, and the coil is driven by program instructions.

Every I/O point of PLC is a certain physical point. The CPU 224 host has 14 digital input endpoints: I0.0–I0.7 and I1.0–I1.5, and 10 digital output endpoints: Q0.0–Q0.7, Q1.0, Q1.1.

(2) Variable from memory (V). V is used to store intermediate results and other data. The CPU 224 host has 5 KB storage bytes: VB0.0–VB5119.7. It can be used in bit, byte, word or double word.

(3) Internal Mark bit storage area (M). M is used as a control relay (also called an intermediate relay) to store intermediate operands or other control information. The addressing range of S7-200 PLC is M0.0–M31.7, and the data in the storage area can be accessed by bit, byte, word or double word.

(4) Sequence control relay storage area (S). S is also called status element to realize sequence control and step control. The addressing range of S7-200 PLC is S0.0–S31.7, and data can be accessed by bit, byte, word or double word.

(5) Special Mark bit memory (SM). The addressing range of CPU 224 host is SM0.0–SM179.7, with a total of 180 bytes. Among them, the 30 bytes of SM0.0–SM29.7 are read-only areas.

① SMB0 is the status bit in byte, which is updated by the S7-200 CPU at the end of each scan cycle. It is defined as follows.

SM0.0: RUN status monitoring. If PLC is in RUN status, this status bit is always 1.

SM0.1: the statue bit is 1 at the first scan, and when the PLC changes from STOP status to RUN status, it outputs an ON signal for one scan cycle, which is used for program initialization.

SM0.2: When the data in RAM is lost, it outputs an ON signal for one scan cycle for error handling.

SM0.3: PLC enters RUN statue after power-on, and it outputs an ON signal for one scan cycle.

SM0.4: minute pulse. This bit outputs a divided clock pulse with a 50% duty cycle for time reference or simple delay.

SM0.5: second pulse. This bit outputs a second clock pulse with a 50% duty cycle. It can be used for time benchmarking.

SM0.6: scan clock. For one scan cycle, it is in ON (high level)status, and for the other scan cycle, it is OFF (low level) in alternate mode.

SM0.7: work mode switch position indication. If it is 0, the switch is in TERM position; if it is 1, the switch is in RUN position. When it is 1, the free port communication mode is available.

② SMB1 is the instruction status bit byte, which is often used in tables and mathematical operations. Some bits are defined as follows.

SM1.0: zero flag. When the operation result is 0, this bit is 1.

SM1.1: overflow flag. This bit is set to 1 when the operation result overflows or an illegal value is detected.

SM1.2: negative number sign. When the result of mathematical operation is negative, this bit is 1.

(6) Local memory (L). S7-200 has 64 bytes of local memory, the addressing range is LB0.0-LB63.7, of which 60 bytes can be used as temporary memory or to pass parameters to subroutines, and the last 4 bytes are reserved for the system.

(7) Timer (its function is equivalent to time relay). The timer in S7-200 CPU is a device that accumulates time increments to the internal clock and is used for time control. The addressing range is T0–T255(22X); T0–T127(21X).

(8) Counter. It is mainly used to accumulate the number of input pulses. There are a 16-bit preset value and a current value register, and a 1-bit status bit. The current value register is used to accumulate the number of pulses. When the current value of the counter is greater than or equal to the preset value, the status bit is 1. S7-200 CPU provides three types of counters: counting up, counting down, and up/down counting. The addressing range is C0–C255 (22X), C0–C127 (21X).

(9) Analog input/analog output image register (AI/AQ). The analog quantity input circuit of S7-200 converts the external input analog quantity (such as temperature, voltage) into a digital quantity with a word length (16 bits), and stores it in the analog quantity input image register area.

The AI addressing range is AIW0, AIW2, ... , AIW62, the starting address is defined as an even-digit address, and there are 32 analog quantity input points in total.

The S7-200 analog quantity output circuit is used to convert a word-length (16-bit) digital value in the analog quantity output image register area into an analog current or voltage output.

The addressing range of AQ is AQW0, AQW2,..., AQW62, the starting address also adopts even-digit address, and there are 32 analog quantity output points in total.

(10) Accumulator (AC). It is used to temporarily store data. S7-200 PLC provides four 32-bit accumulators AC0–AC3. The accumulator supports byte (B), word (W) and double word (D) access.

(11) High-speed counter (HC). CPU22X series PLC provides 6 high-speed counters HC0, HC1, ..., HC5 (the highest frequency of each counter is 30 kHz) to accumulate events faster than the CPU scan rate. The current value of the high-speed counter is a double-word signal integer.

✍ Practice and operation

1. Task preparations

According to the requirements of this training task, prepare the list of components that need to be used as shown in Table 4-4.

Table 4-4 Component preparation list

No.	Name	Specification model	Amount
1	Programmable controllers	CPU226	1
2	Automatic feeding and loading control system unit	YL360-02	1
3	Download cable	PPI cable	1
4	Connection line	YL360 dedicated training platform	Appropriate amount
5	Computer	Self-provided in the training room	1

2. Training content

The control panel of the automatic feeding and loading control system is shown in Figure 4-1. The control requirements are as follows: In the initial state, the red light L2 is off and the green light L1 is on, indicating that the car is allowed to enter and load. Hopper K2, motors M1, M2, and M3 are all OFF. When the car arrives (indicated by the turn-on of S2 switch), L2 is on, L1 is off, and M3 runs. Motor M2 runs after M3 is turned on for 2 seconds, motor M1 runs after M2 starts for 2 seconds, and after a delay of 2 seconds, the hopper K2 opens to discharge. When the car is full (indicated by S2 disconnection), the hopper K2 is closed, the motor M1 stops after a delay of 2 seconds, M2 stops after M1 stops for 2 seconds, and M3 stops after M2 stops for 2 seconds. When L1 is on and L2 is off, it indicates that the car can be driven away.

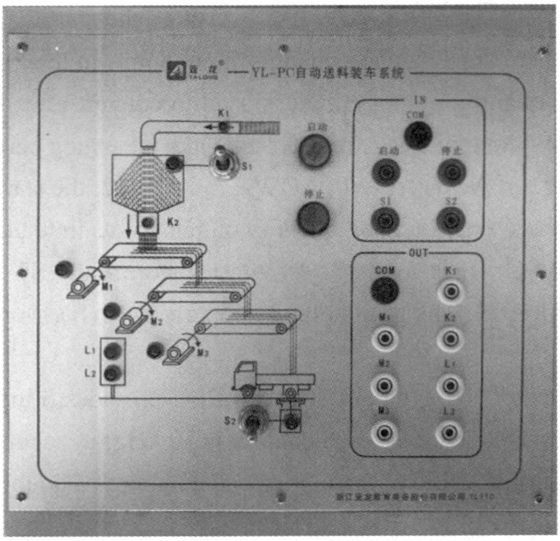

Figure 4-1　Control panel of automatic feeding and loading control system

S1 is the material level detection switch in the hopper. When S1 is closed, it means that the material is full, and K2 is open. When S1 is disconnected, it means that the hopper is not full, K1 is open, and K2 is not open.

After press the STOP button, the motors M1, M2, M3 stop running. Need to press the RUN button again to start the motors.

The training content of automatic feeding and loading control system is arranged as follows.

(1) According to the components of the automatic feeding and loading control system module, compile the I/O allocation table of the automatic feeding and loading control system as shown in Table 4-5;

(2) Carry out system wiring according to the PLC control schematic diagram of the automatic feeding and loading control system (as shown in the Figure 4-2);

(3) Compile the PLC control program of the automatic feeding and loading control system;

(4) Download the control program, and debug and run the program.

Table 4-5　I/O allocation table of automatic feeding and loading control system

Input		Output	
I0.0	Hopper upper limit switch S1	Q0.0	Feeding switch K1
I0.1	Position detection switch S2	Q0.1	Hopper switch K2
		Q0.2	Motor M1
		Q0.3	Motor M2
		Q0.4	Motor M3
		Q0.5	Green light L1
		Q0.6	Red light L2

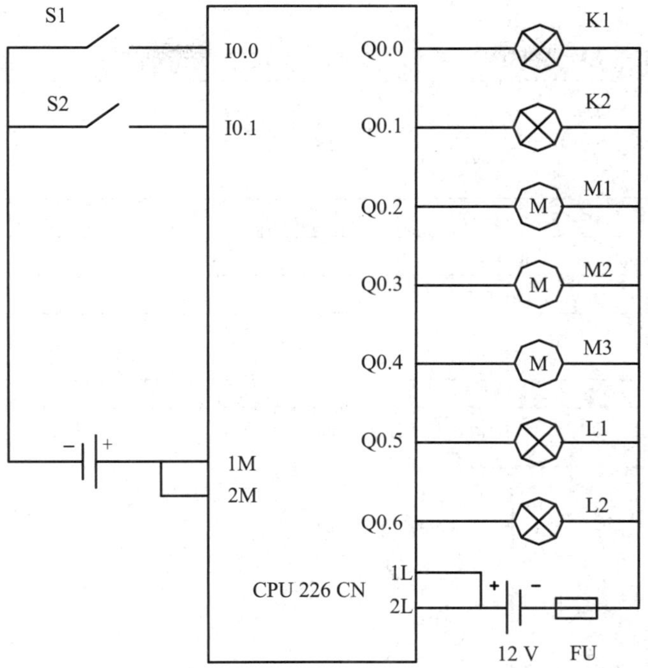

Figure 4-2 PLC control schematic diagram

The PLC control program of the automatic feeding and loading control system is shown in Figure 4-3 below.

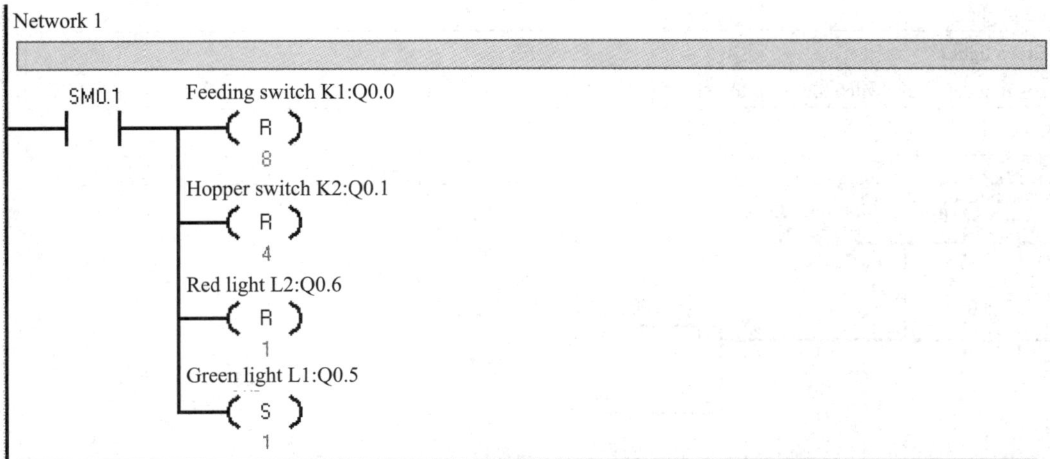

Symbol	Address	Comment
Red light L2	Q0.6	
Hopper switch K2	Q0.1	
Green light L1	Q0.5	
Feeding switch K1	Q0.0	

Network 2

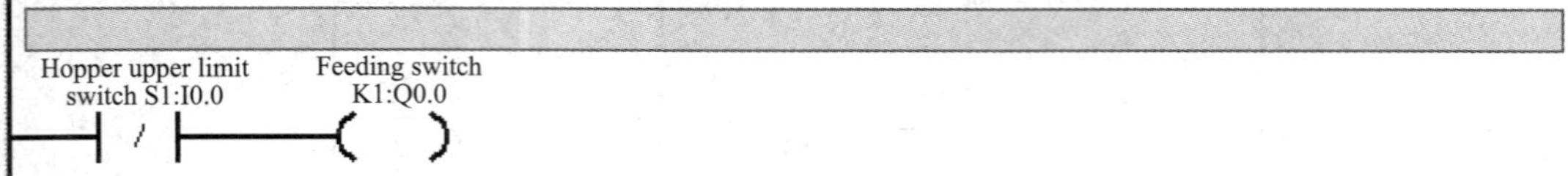

Symbol	Address	Comment
Hopper upper limit switch S1	I0.0	
Feeding switch K1	Q0.0	

Network 3

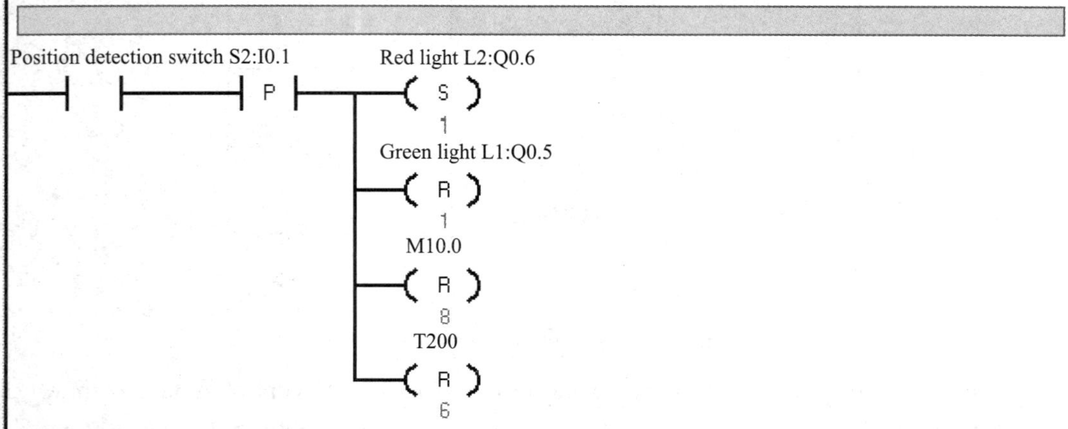

Symbol	Address	Comment
Red light L2	Q0.6	
Green light L1	Q0.5	
Position detection switch S2	I0.1	

Network 4

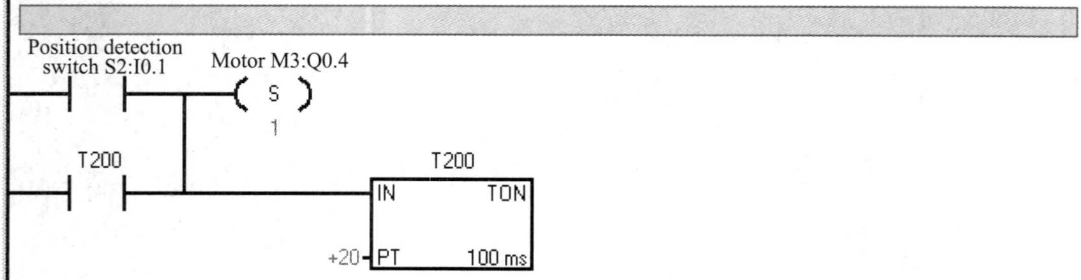

Symbol	Address	Comment
Motor M3	Q0.4	
Position detection switch S2	I0.1	

Network 5

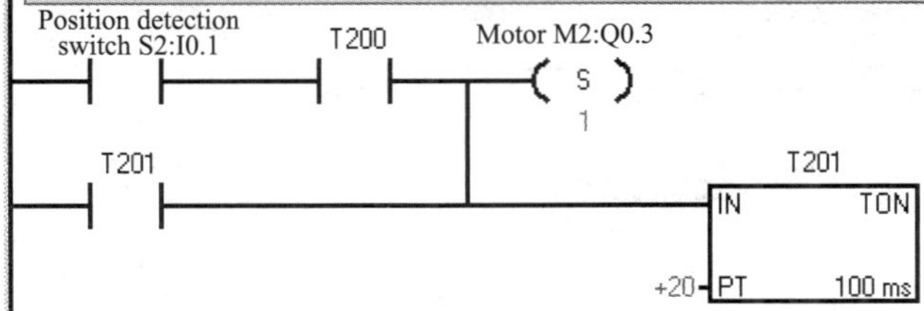

Symbol	Addres	Comment
Motor M2	Q0.3	
Position detection switch S2	I0.1	

Network 6

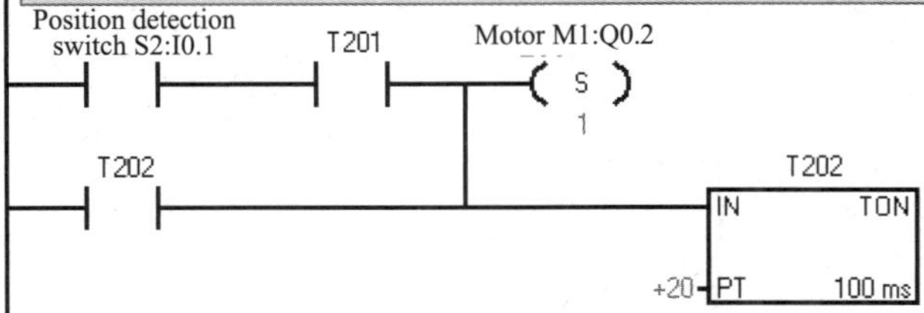

Symbol	Addres	Comment
Motor M1	Q0.2	
Position detection switch S2	I0.1	

Network 7

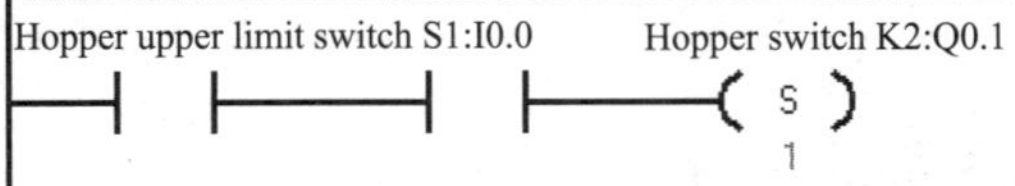

Symbol	Addres	Comment
Hopper switch K2	Q0.1	
Hopper upper limit switch S1	I0.0	

Network 8

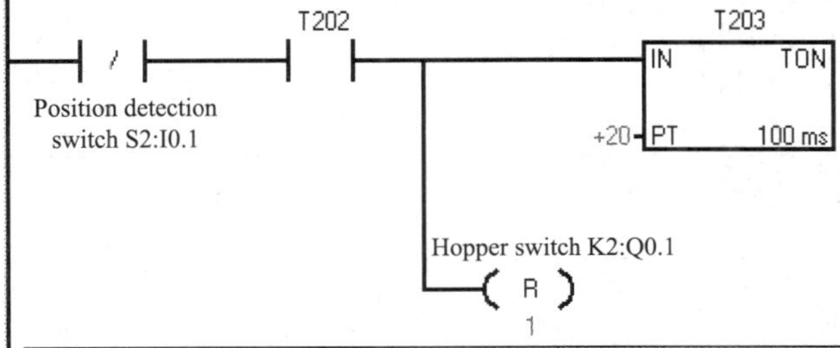

Symbol	Addres	Comment
Hopper switch K2	Q0.1	
Position detection switch S2	I0.1	

Network 9

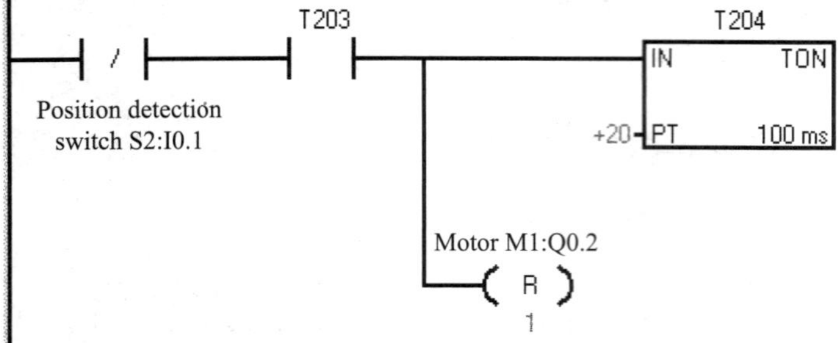

Symbol	Addres	Comment
Motor M1	Q0.2	
Position detection switch S2	I0.1	

Network 10

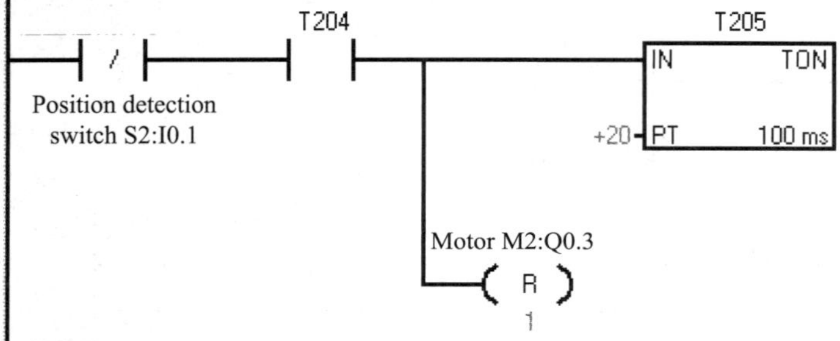

Symbol	Addres	Comment
Motor M2	Q0.3	
Position detection switch S2	I0.1	

Network 11

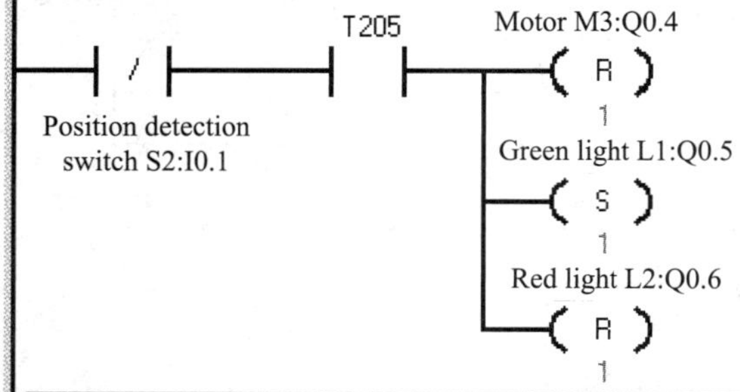

Symbol	Addres	Comment
Motor M3	Q0.4	
Red light	Q0.6	
Green light	Q0.5	
Position detection switch S2	I0.1	

Figure 4-3 PLC program of automatic feeding and loading control system

3. Task procedures

Task procedures are as follows.

(1) First turn off the power switch of the PLC host, connect the wires strictly as shown in Figure 4-2. Pay attention to that the positive and negative of the 24 V power supply should not be in short-circuited, otherwise, the PLC contacts will be damaged.

(2) Insert the power cord into the power hole on the surface of the PLC host, and then plug the other end into the 220V power strip.

(3) Put the power of the PLC host in the ON state, and put the PLC serial port in the STOP state, download the program to the PLC through the computer or programmer, and then put the PLC serial port in the RUN state after downloading.

(4) Follow the steps as follows.

① After start the system, the green light L1 is on and the hopper K1 light is on.

② Turn on the switch S2 (the indicator light is on), the green light L1 is off, the red light L2 is on, and the motors M3, M2, and M1 light up in sequence.

③ Turn on the detection switch S1 (the indicator light is on), the hopper K1 is off, and the hopper K2 is on.

④ Pull down the detection switch S2 (the indicator light is on), the motors M1, M2, and M3 are turned off in turn, L1 is on, and the hopper K1 is on (returning to the initial state).

4. Precautions

Precautions are as follows:

(1) Do not touch the electrified components, and strictly follow the operating sequence of "wiring first, then power on; first connect the control circuit, then connect the power supply; disconnect the power first, then disconnect the wires".

(2) When wiring, you must connect the load terminal first, then the power terminal; connect the ground terminal first, and then the power cord.

(3) If abnormal phenomena (such as noise, fever, burnt smell) are found, the power supply should be cut off immediately, the scene should be kept, and they should be reported to the instructor.

(4) The wiring must be done corresponding to the electrical schematic diagram, especially the power wiring must be connected correctly, which will cause the power module to burn out and damage the module.

(5) When downloading the program, if you encounter failure information, such as not being able to find the CPU, you should check whether the download cable is connected properly.

5. Notes and summary

___

___

___

___

Task 2 Programming and debugging of traffic light automatic control system

✍ Relevant knowledge

1. Basic bit operation instructions

The format of basic bit operation instructions is shown in Table 4-6.

Table 4-6 Format of basic bit operation instructions

LAD	STL	Description
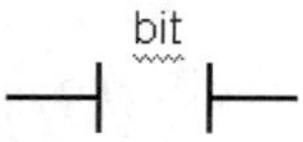	LD BIT LDN BIT	Normally-open-contact/normally-closed-contact for the start of the network segment
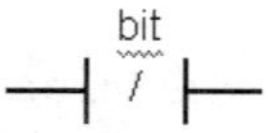	A BIT AN BIT	Normally-open-contact/normally-closed-contact in series, logical "AND"/"NAND" instructions
	O BIT ON BIT	Normally-open-contact/normally-closed-contact in parallel, logical "OR"/"NOR" instructions
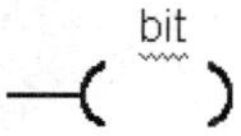	= BIT	Coil output, logic position set instruction

(1) Basic bit operation instruction operand addressing range: I, Q, M, SM, T, C, V, S, L, etc.

(2) Instruction mnemonic: LD (Load), LDN (Load Not), A (And), AN (And Not), O (Or), ON (Or Not), = is coil output setting instruction.

(3) Block operation instruction "or" at the format of OLD (with no operating components involved). The block operation instruction "or" is to connect two adjacent circuit blocks starting with LD in the ladder diagram in parallel.

(4) Block operation instruction "and" at the format of ALD (with no operating components involved). Block operation instruction "and" is to connect two adjacent circuit blocks starting with LD in the ladder diagram in series.

2. Negation and null operation instructions

The format of negation and null operation instructions is shown in Table 4-7.

Table 4-7 The format of negation and null operation instructions

LAD	STL	Description
	NOT	Negation instruction
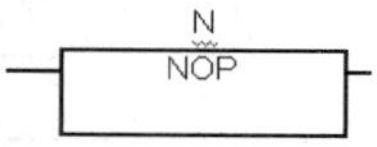	NOP N	Null operation instruction

3. Set/reset instructions

The format of the set and reset commands is shown in Table 4-8.

Table 4-8 Format of set and reset commands

LAD	STL	Description
S-bit —(S) N	S S-BIT, N	N components from the start bit are set to 1
S-bit —(R) N	R S-BIT, N	N components from the start bit are cleared to 0

4. Edge trigger instruction

Edge trigger instruction refers to the use of edge trigger signals to generate scan pulses for one machine cycle, usually used for pulse shaping. The format of edge trigger instruction is divided into two categories: positive transition trigger (Edge Up, EU) and negative transition trigger (Edge Down, ED). Positive transition trigger refers to that the rising edge of the input pulse makes the contact ON for one scan period. Negative transition trigger refers to that the falling edge of the input pulse makes the contact ON for one scan period. These two formats are with no operating components involved.

An example of an edge trigger instruction program is shown in Figure 4-4.

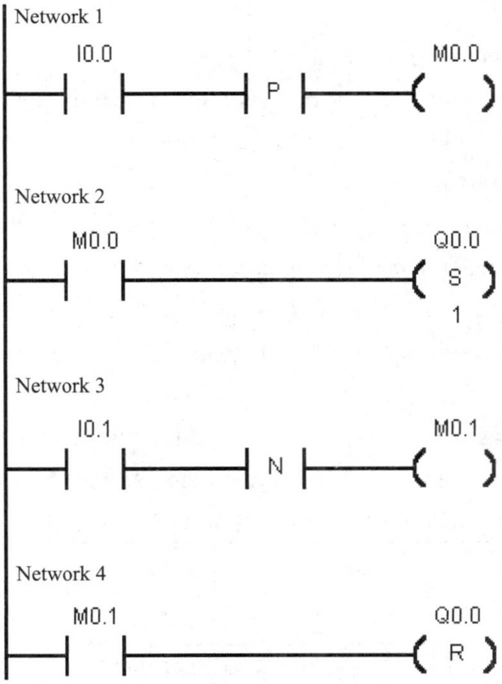

Figure 4-4 An example of an edge trigger instruction program

The timing sequence analysis of edge trigger instruction program is shown in Figure 4-5.

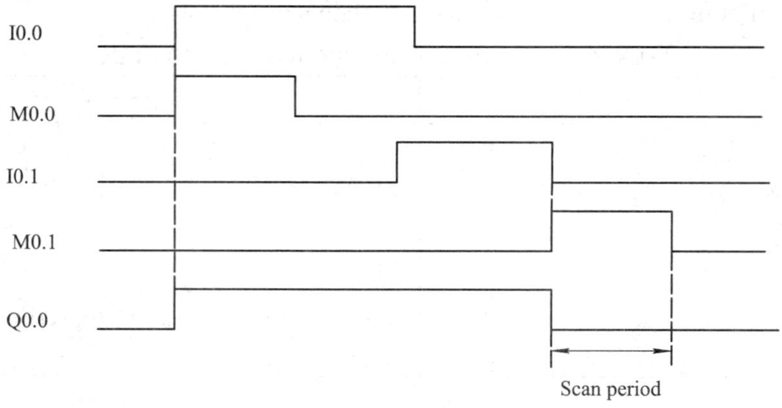

Figure 4-5 Timing sequence analysis diagram of edge trigger instruction program

(1) On the positive transition trigger of I0.0, EU generates a scan period clock pulse, the M0.0 coil is energized, and the normally-open-contact of M0.0 is closed for a scan period, which makes the trigger command of output coil Q0.0 valid (output coil Q0.0 =1) and keep.

(2) On the negative transition trigger of I0.0, ED generates a scan period clock pulse driving the output coil M0.1 to energize for a scan period, the normally-open-contact of M0.1 is closed, which makes the resetting command of output coil Q0.0 is in working (Q0.0=0) and kept.

5. Timer

1) Classification of working methods

According to working methods, timers are divided into power-on delay type (TON), power-on delay type recall (hold type) (TONR), and power-off delay type (TOF).

2) Classification of time base

According to time base standard, timers resolution are divided into 1 ms, 10 ms and 100 ms. time base standards, the refresh mode of timing accuracy and timing range is different.

Calculation of timing time: T=PT×S. T is the true timing time, PT is the preset value of timing time, and S is the resolution level. For example: TON instruction uses timer T37, the preset value of timing time is 150, then the true timing time is T=150×100=15 000 ms=15 s.

The larger the time base, the longer the timing time, but the worse the resolution. The time base classification is shown in Table 4-9. The timer command format is shown in Table 4-10.

Table 4-9 Time base classification

Working mode	Timer resolution/ms	Maximum timing range/s	Timer number
TONR	1	32.767	T0，T64
	10	327.67	T1-T4，T65-T68
	100	3276.7	T5-T31，T69-T95

Continued table

Working mode	Timer resolution/ms	Maximum timing range/s	Timer number
TON/TOF	1	32.767	T32，T96
	10	327.67	T33–T36，T97–T100
	100	3276.7	T37–T63，T101–T255

Table 4-10 Timer command format

LAD	STL	Description
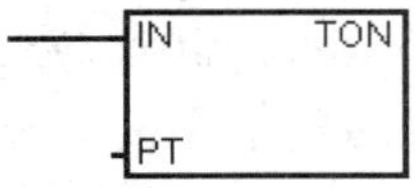	TON	on-delay timer type
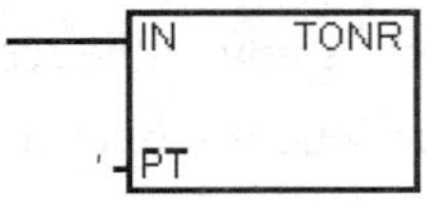	TONR	on-delay timer type with recall
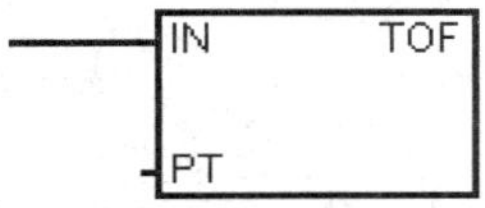	TOF	off-delay timer type

In Table 4-10, IN is the enable input terminal; programming range is T0–T255. PT is the preset value input terminal, the maximum preset value is 32 767; PT data type is INT(integer).

3) On-delay timer

On-delay timer command is used for timing a single interval. During the first on cycle or the first scan period, the timer bit is OFF, and the current value is 0. When the enable input is ON, the timer bit is OFF, and the current value starts counting from 0. When the current value reaches the preset value, the timer bit is ON, and the current value continuously counts to 32 767. The enable input is OFF, the timer is automatically reset, that is, the timer bit is OFF, and the current value is 0.

Command format: TON Txxx, PT

Example: TON T120, 8

Figure 4-6 is an application example of the on-delay timer.

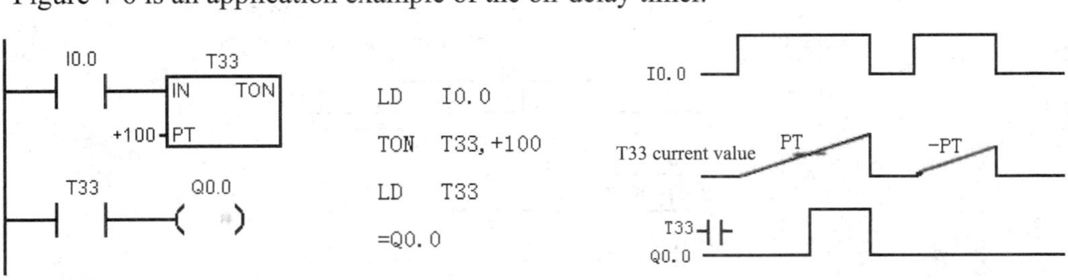

(a) Ladder diagram (b) Instruction statement list (c) Timing sequence analysis diagram

Figure 4-6 Application of on-delay timer

In the example shown in Figure 4-6, when I0.0 is ON, T33 is driven to start timing (counting time base pulses); when the timing reaches the set value PT, the T33 status bit is set to 1, and its normally-open-contact is ON, driving Q0.0 to have output; after that, the current value still increases, but it does not affect the status bit. When I0.0 is OFF, T33 is reset, the current value is cleared to 0, and the status bit is also cleared to 0, that is, it is restored to original statue. If I0.0 is OFF before the set value is reached, T33 will reset follow ingly, and Q0.0 will not have output.

4) On-delay timer with recall

Instruction of on-delay timer with recall (TONR) is used to accumulate timing for many intervals. During the power-on cycle or the first scan, the timer bit is OFF and the current value remains the currents statue. When the enable input is turned on, the timer bit is ON, and the current value starts counting time from 0. The enable input is disconnected, and the timer bit and current value remain the last statue. When the enable input is turned on again, the current value will continue to count from the last held value. When the accumulated current value reaches the preset value, the timer bit will be ON, and the current value will continue to count to 32 767. TONR timer can only be reset by reset command.

Command format: TONR　Txxx, PT

Example: TONR　T20, 63

For the on-delay timer with recall, when the input (IN) is "1", the timer counts time (counting up the time base pulse); when IN is "0", its current value remains (not reset as TON). Next time, When IN is 1, the current value of Txxx increases based on the original holding value, and the current value is in comparison with the set value PT: when the current value is greater than or equal to the set value, the timer status bit is set to "1", In the future, even if IN is "0" again, the timer will not reset. To reset the timer, a reset command must be used. The application of on-delay timer with recall is shown in Figure 4-7.

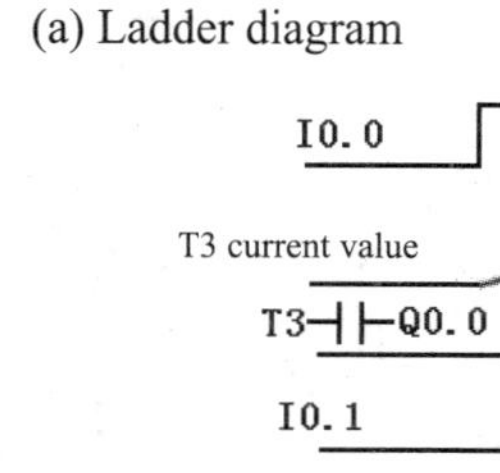

(a) Ladder diagram

```
LD    I0.0
TONR  T3, +100
LD    T3
=     Q0.0
LD    I0.1
R     T3, 1
```

(b) Instruction statement list

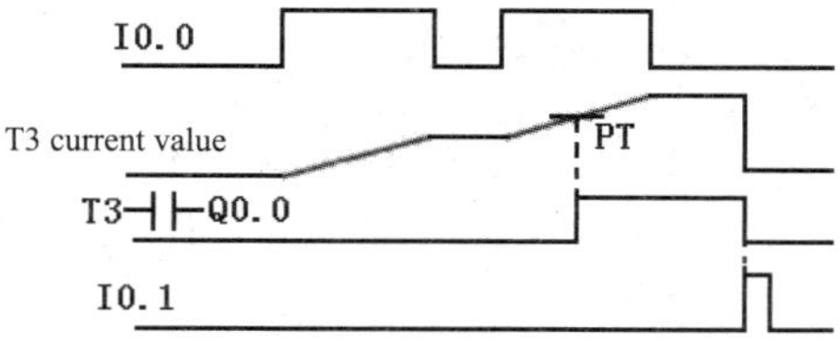

(c) Timing diagram

Figure 4-7　Application of on-delay timer with recall

5) Off-delay timer

Off-delay timer instruction is used to accumulate timing for single interval after disconnection. During the first power-on cycle or the first scan, the timer bit is OFF, and the current value is 0. When the enable input is ON, the timer bit is ON, and the current value is 0. When the enable input changes from ON to OFF, the timer starts counting. When the current value reaches the preset value, the timer bit is OFF, the current value is equal to the preset value, and the counting stops. After TOF reset, if the enable input has a negative transition from ON to OFF, timer can be restarted.

Command format: TOF Txxx, PT

Example: TOF T35, 6

6. Counter

The counter is used to accumulate the number of input pulses. The counter is composed of integrated circuits and is a very widely used programming element, often used to count the number of products.

There are three counter instructions: count up instruction CTU, count up and down instruction CTUD, and count down instruction CTD.

The instruction operand has 4 parts: number, preset value, pulse input and reset input.

1) Count up

Count up instruction CTU. In the first scan, the timer bit is OFF, and the current value is 0. At each rising edge of pulse input, the counter counts 1 time and the current value increases by 1 unit. When the current value reaches the preset value, the counter bit is ON, and the current value continues to count to 32 767, then it stops counting. The reset input is valid, in other words, the reset instruction is executed, the counter is automatically reset, that is, the counter bit is OFF, and the current value is 0.

Command format: CTU Cxxx, PV

Example: CTU C20, 3

Figure 4-8 is an example of a count up program.

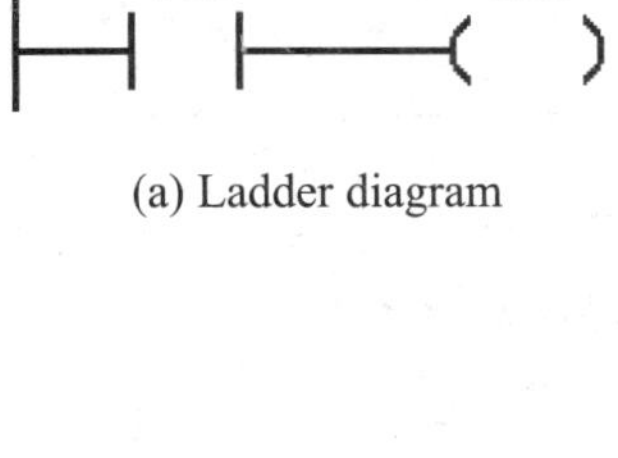

(a) Ladder diagram

```
LD    I0.0
LD    I0.1
CTU   C10, +5
LD    C10
=     Q0.0
```

(b) Instruction statement list

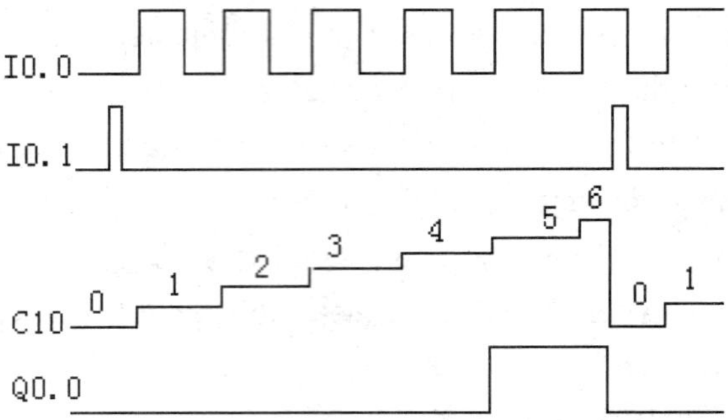

(c) Timing diagram

Figure 4-8　Example of count up program

2) Count up and down

In count up and down instruction CTUD, there are two pulse input terminals: the CU input terminal used for up counting and the CD input terminal used for down counting.

Instruction format: CTUD　Cxxx, PV

Example: CTUD　C30, 5

An example of count up and down program is shown in Figure 4-9.

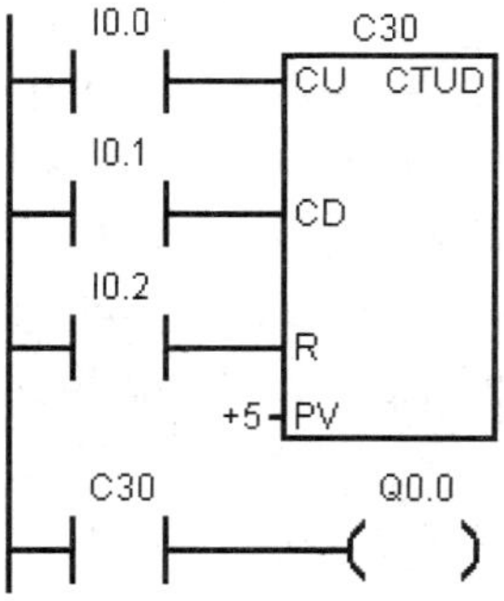

(a) Ladder diagram

```
LD      I0. 0
LD      I0. 1
LD      I0. 2
CTUD    C30, +5
LD      C30
=       Q0. 0
```

(b) Instruction statement list

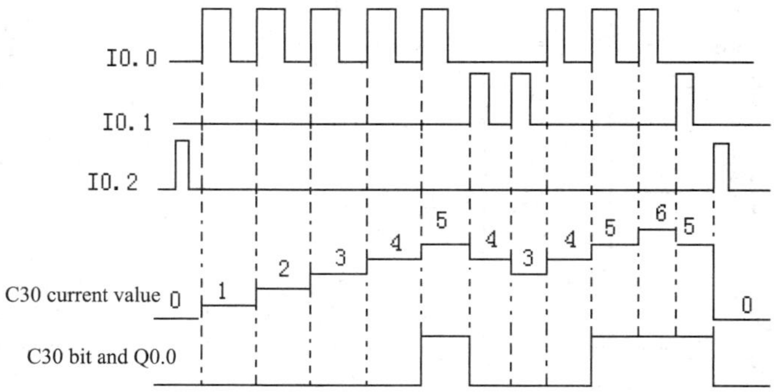

(c) Timing diagram

Figure 4-9　Example of count up and down program

3) Count down

In count down instruction CTD, the pulse input CD is used to count down. In the first scan, the timer bit is OFF, and the current value is equal to the preset value PV. When the counter detects each rising edge of CD input, the current value of the counter decreases by 1 unit, and when the current value is decreased to 0, the counter bit is ON.

The reset input is valid, in other words the reset instruction is executed, the counter is automatically reset, that is, the counter bit is OFF, and the current value reset to the preset value instead of 0.

Command format: CTD Cxxx, PV

Example: CTD C40, 4

An example of count down program is shown in Figure 4-10.

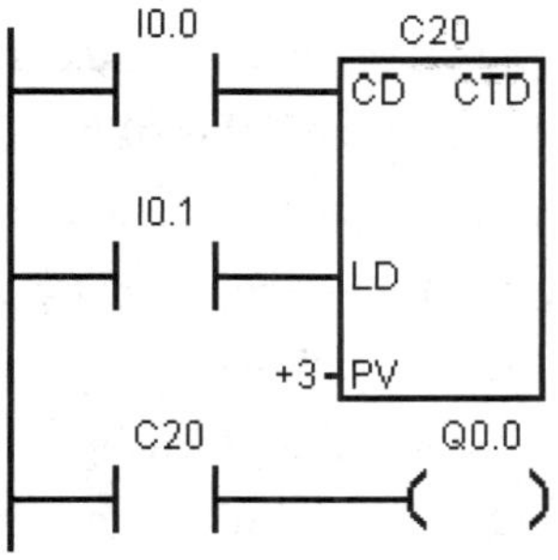

(a) Ladder diagram

```
LD      I0.0
LD      I0.1
CTD     C20, 3
LD      C20
=Q0.0
```

(b) Instruction statement list

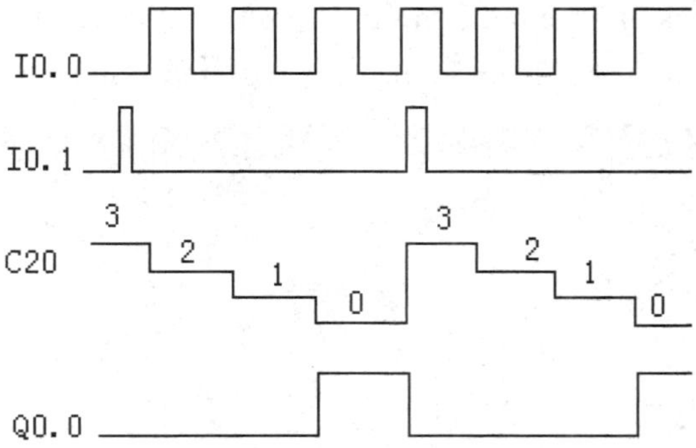

(c) Timing diagram

Figure 4-10 Example of count down program

✍ Practice and operation

1. Task preparations

According to the requirements of this training task, prepare the components as in Table 4-11.

Table 4-11　Component List

No.	Name	Specification model	Amount
1	Programmable controllers	CPU226	1
2	Traffic light automatic control system unit	YL360-03	1
3	Download cable	PPI cable	1
4	Connection line	YL360 dedicated for training platform	Appropriate amount
5	Computer	Self-provided training room	1

2. Training content

The traffic light automatic control system unit panel is shown in Figure 4-11. The requirements for traffic light automatic control system design are as follows.

(1) Traffic light automatic control system unit is equipped with start and stop switches S1 and S2 to control the start and stop of the system. S3 can shield the lights of traffic lights.

(2) Traffic light display mode. When the red light in the east-west direction is on, the green light in the north-south direction is on. When the green light is going to change, the green light flashes three times with flashing period 1 second, and then the yellow light is on for 2 seconds. When the yellow light in the north-south direction goes dark, the green light in the east-west direction is on and the red light in the north-south direction is on. When the green light in the east-west direction is going to change, the green light flashes three times with flashing period 1 second, and then the yellow light at this direction is on for 2 seconds. When the yellow light in the east-west direction is off, the red light in the east-west direction is on, and the green light in the north-south direction is on... The routine goes repeatedly and continuously.

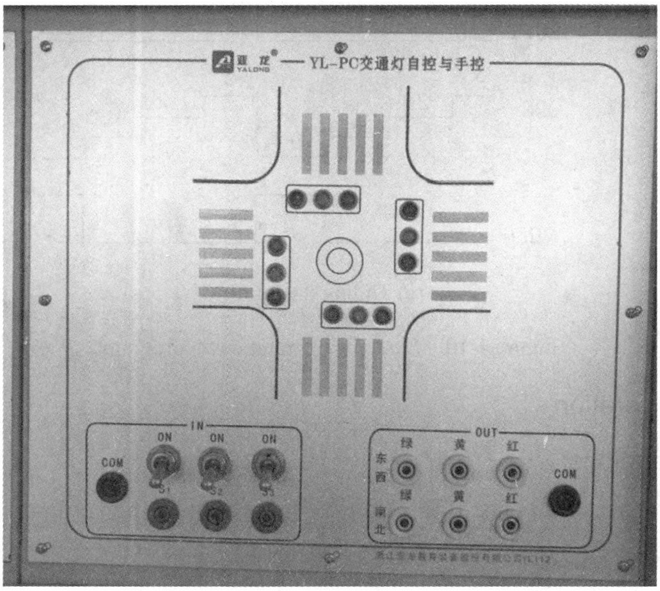

Figure 4-11　Traffic light automatic control system unit panel

The training content of traffic light automatic control system is arranged as follows:

(1) According to the components of the traffic light automatic control system unit, compile the traffic light automatic control system I/O allocation as shown in Table 4-12;

(2) Carry out the wiring according to the PLC control schematic diagram of the traffic light control automatic system (as shown in the Figure 4-12);

(3) Compile the PLC control program of the traffic light automatic control system;

(4) Download the control program, then debug and run the program.

Table 4-12 I/O allocation table of traffic light automatic control system

Input		Output	
I0.0	Start switch S1	Q0.0	East-west direction red light
I0.1	Stop switch S2	Q0.1	East-west direction green light
I0.2	Shield switch S3	Q0.2	East-west direction yellow light
		Q0.3	North-south direction red light
		Q0.4	North-south direction green light
		Q0.5	North-south direction yellow light

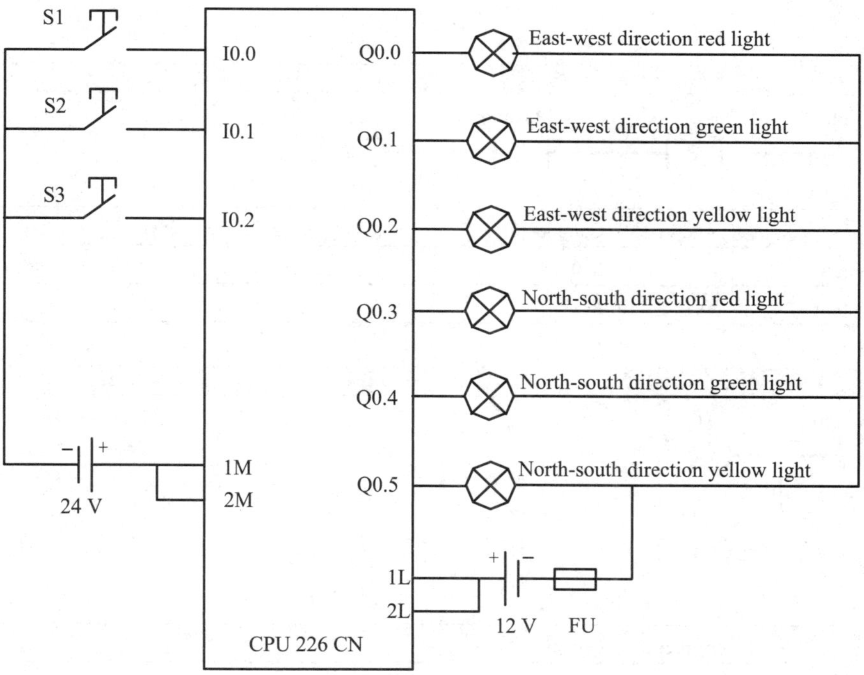

Figure 4-12 PLC control schematic diagram of the traffic light automatic control system

3. Task procedures

Task procedures are as follows:

(1) First put the power switch of the PLC host in OFF/dial-down state, and wiring strictly as shown in Figure 4-12. Note that the positive and negative of the 24 V power supply should not be short-circuited, otherwise, the PLC contacts will be damaged.

(2) Insert the power cord into the power hole on the surface of the PLC host, and then plug the other end into the 220 V power strip.

(3) Put the power switch of the PLC host in ON/dial-up state, and the PLC serial port must be placed in the STOP state, and then download the program to the PLC through the computer or programmer. After downloading, put the PLC serial port in RUN state.

PLC program diagram of the traffic light automatic control system is shown as Figure 4-13.

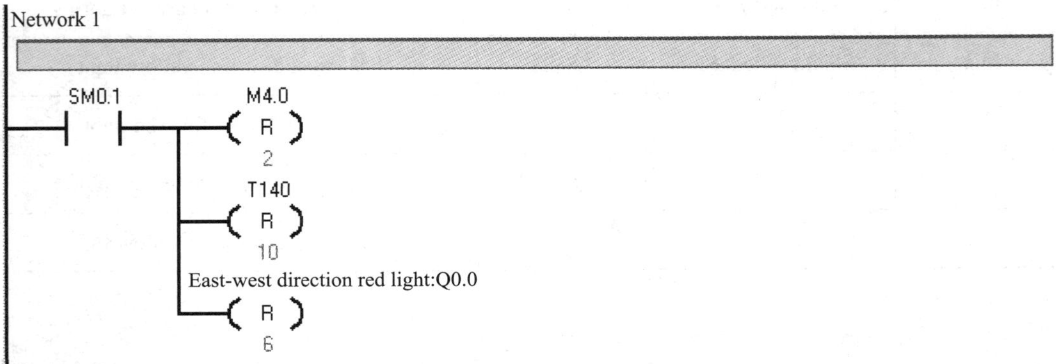

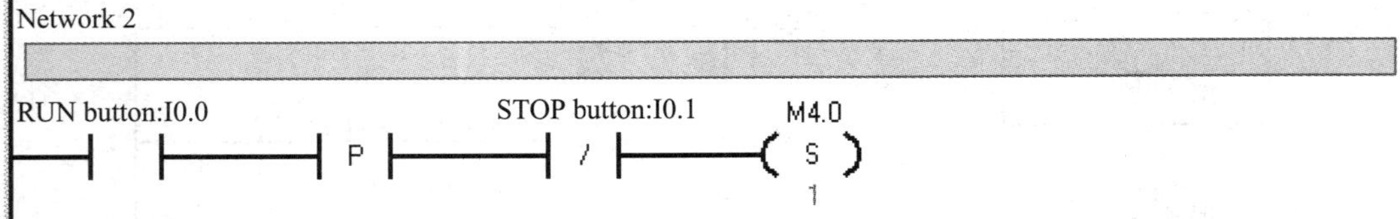

Symbol	Address	Comment
RUN button	I0.0	I0.0
STOP button	I0.1	I0.1

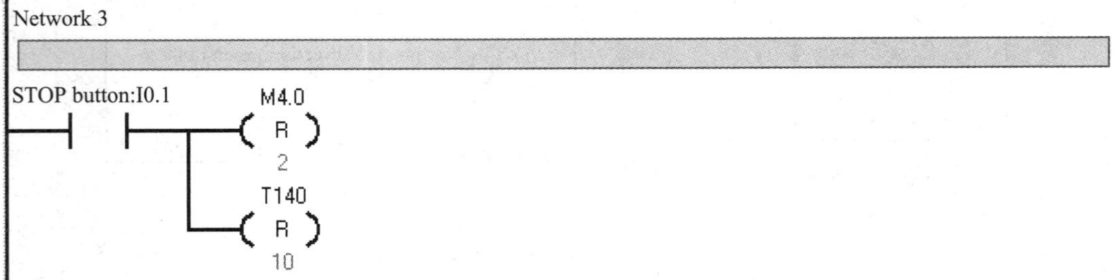

Symbol	Address	Comment
STOP button	I0.1	I0.1

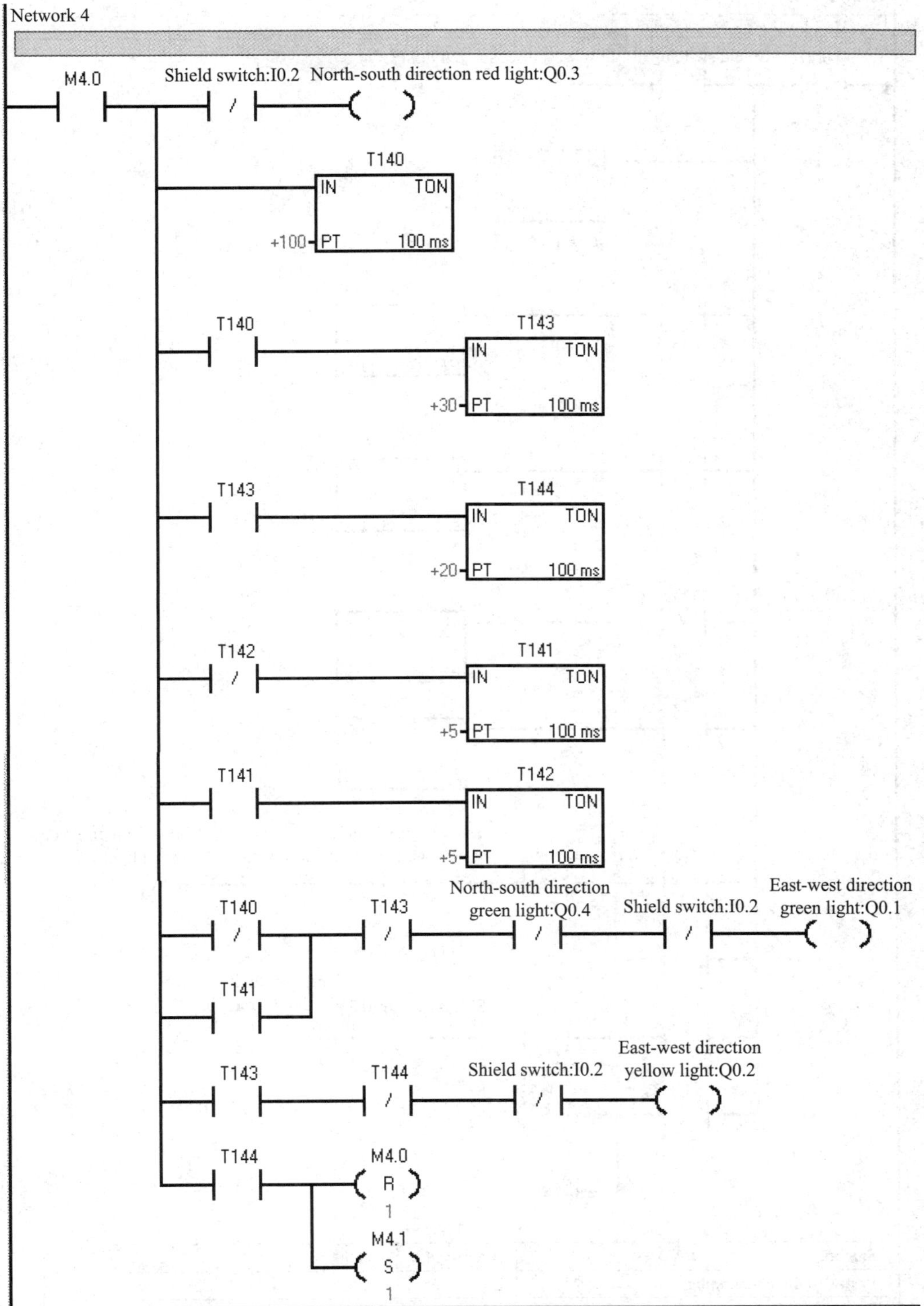

Symbol	Address	Comment
East-west direction yellow light	Q0.2	Q0.2
East-west direction green light	Q0.1	Q0.1
North-south direction red light	Q0.3	Q0.3
North-south direction green light	Q0.4	Q0.4
Shield switch	I0.2	I0.2

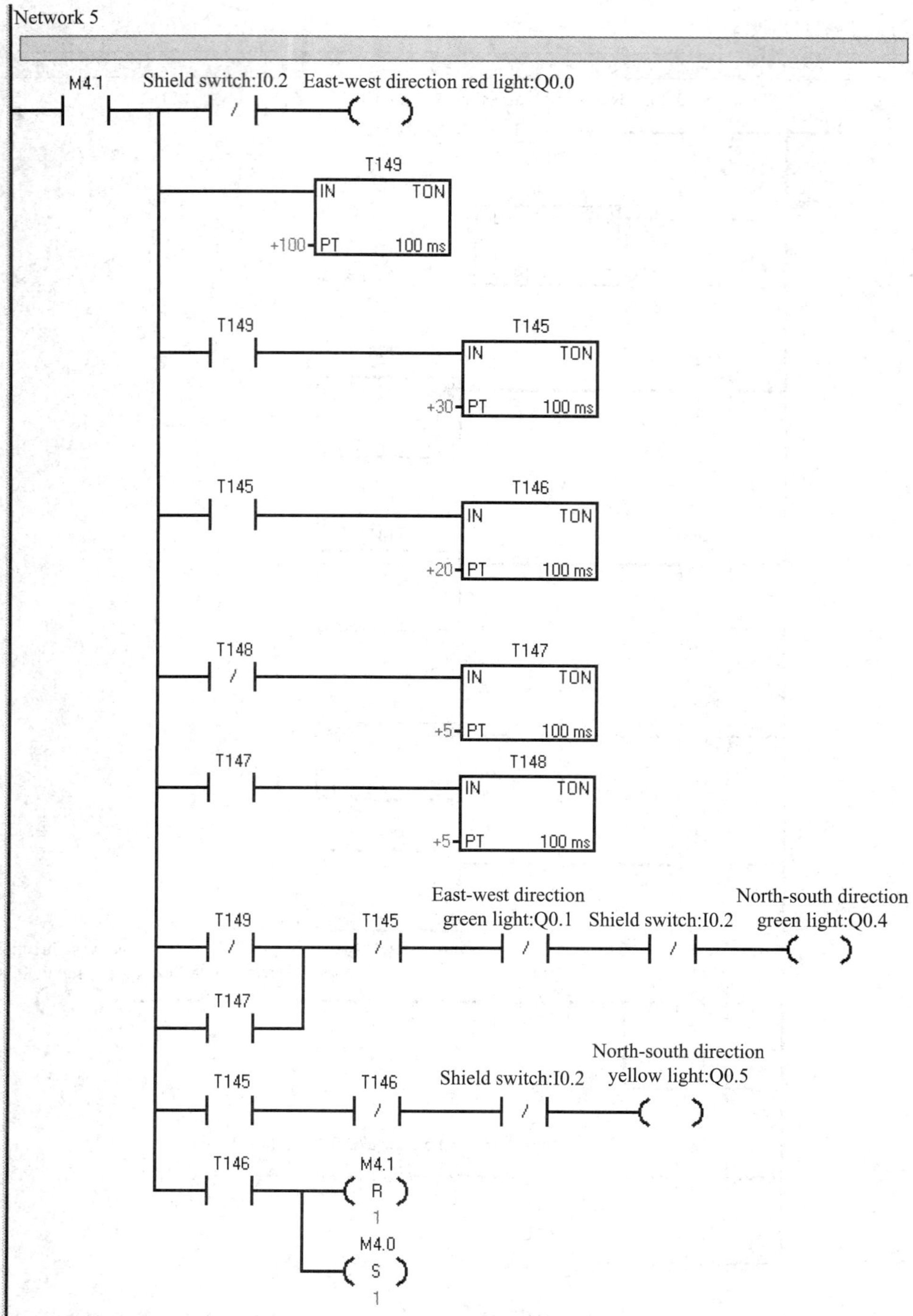

Symbol	Address	Comment
East-west direction red light	Q0.0	Q0.0
East-west direction green light	Q0.1	Q0.1
North-south direction yellow light	Q0.5	Q0.5
North-south direction green light	Q0.4	Q0.4
Shield switch	I0.2	I0.2

Figure 4-13 PLC program diagram of traffic light automatic control system

(4) Training operation processes are as follows:

① Dial up and then dial down to start S1, and observe the changes of traffic lights.

② Dial up the shield switch S3 to observe the change of the light, dial down S3 to observe the change of the light.

③ Dial up the stop switch S2 to observe the change of the light, dial down S2 to observe the change of the light.

④ Compare the functions of switches S2 and S3. S2 can make the light go dark forever, and S3 can make the light go dark temporarily.

4. Precautions

Precautions are as follows:

(1) Do not touch the live parts, and strictly follow the operating sequence of "wiring first, then power on; first connect the control circuit, then connect the power supply; disconnect the power first, then disconnect the wirings".

(2) When wiring, you must connect the load terminal first, then the power terminal; connect the ground terminal first, and then the power cord.

(3) If abnormal phenomena (such as noise, fever, burnt smell) are found, the power supply must be cut off immediately; the scene must be kept, and the instructor must be reported.

(4) The wiring must correspond to the electrical schematic diagram, especially the power wiring must not be connected incorrectly, which will cause the power unit to burn out and damage the module.

(5) When downloading the program, if you encounter failure information such as not being able to find the CPU, you should check whether the download cable is connected properly.

5. Notes and summary

____

____

____

____

____

Task 3 Programming and debugging of four-floor elevator control system

✍ Relevant knowledge

1. Local variable table

1) Definition of local variables

With regard to the definition of local variables, we can introduce its definition in comparision with the definition of global variables. The variables defined in the SIMATIC symbol table or IC's global variable table are global variables. Each Program Organizational Unit (POU) in the program has its own local variable table composed of 64-byte L memory. POU are used to define variables with scope limitation (local variables), and local variables are only valid in the POU in which it is created. On the contrary, global variables (symbols) are valid in each POU and can only be defined in the variable table/global variable table. When the global variables and the local variables have the same name, local variables defining in the POU have high priority, and the definition of global variables can only be used in other POUs.

Local variables have the following advantages:

(1) If local variables are used in the subprogram, the subprogram can be transplanted to other projects under the condition that the absolute addresses or global symbols are not used, and no changes are made.

(2) If temporary variables (TEMP) are used in the subroutine, the same piece of physical memory can be reused in different programs. Local variables are also used to transfer input parameters and output parameters between the subroutine and the program invoking this TEMP.

2) Types of local variables

(1) TEMP (temporary variables). TMP is a variable temporarily stored in the local data area. Only when the POU is executed, the defined temporary variables are used. After the POU is executed, the value of the temporary variables is no longer saved. In the main program and interrupt program, there are only TEMP variables in the local variable table.

(2) IN (input variables). IN is the input parameter passed into the subroutine, and provided and invoked by the POU. If the parameter is invoked by direct addressing, such as VB10, the value of the address specified directly is passed into the subroutine. If the parameter is invoked by indirect addressing, such as *ACl, the value of the address specified by the pointer is passed into the subroutine. If the parameter is a constant (such as DW#12345) or an address (such as &VB100), the value of the constant or address is passed into the subroutine.

(3) OUT (output variables). OUT is the execution result of the subroutine, and it is returned to the POU invoking or calling.

(4) IN_OUT (input/output variables). The initial value of IN_OUT is transferred to the subroutine by the POU calling, and the same variable is used to return the execution result of the subroutine to the POU calling. Constants and addresses (such as &VB00) cannot be used as input/output variables.

3) Address allocation of local variables

When assigning values in the local variable table, you only need to specify the type (TEMP, IN, IN_OUT, or OUT) of the local variables and data type, without specifying the memory address. The program editor automatically specifies the memory location for all local variables in the local memory. The starting address is IB0, 1 to 8 consecutive bit parameters are assigned a byte, and the bit address in the byte is Lx07 (x is the byte address). Byte, word, and double word values are allocated in byte order in local memory, such as LBLWx or LDx.

4) Add new variables in the local variable table

For the main program and interrupt program, the local variable table shows a group of rows that have been pre-defined as TEMP. To add a row to the table, just right-click a row in the table, execute the "Insert" → "Row" command in the pop-up menu, and insert a new row above the selected row. Or execute the menu command to insert the "next line" instruction and insert a new line below the selected line. For subroutines, the display data type of the local variable table is pre-defined as the series row of IN, IN_OUT, OUT and TEMP, and their order cannot be changed. If you want to add new local variables, you must right-click an existing row and use the pop-up menu to insert another local variables of the same type above or below the selected row.

5) Local variable data type check

When local variables are passed into subroutine as parameters, the data type specified in the local variable table of the subroutine must match the data type value in the POU calling.

2. Subroutine programming application

The control program of S7-200 CPU is composed of main program (OB1), subroutine and interrupt program. STEP7 Micro/WIN SMART program software provides a separate page for each POU in the program editor window. The main program is always on the first page, followed by subprogram or interrupt program. Because each POU is stored in separate page in the program editor window, the subroutine or interrupt program automatically returns when it is executed to the end, and there is no need to add a return instruction; the conditional return instruction can be used in the subroutine or interrupt program.

1) The role of subroutine

Subroutine is often used in places where the same task needs to be performed repeatedly. You only need to compile the subroutine once, and such subroutine is available to other programs (directly calling it and it is unnecessarily to compile a new one). The calling of the subroutine is conditional, the instructions in the subroutine will not be executed when it is not called, so the use of the subroutine can reduce the scan time. When compiling a complex PC program, it is best to divide all control functions into several that are in line with process control regular sub-function blocks. Each sub-function block is composed of one or more parts of subroutine. The subroutine makes the program structure simple and clear, easy to debug, troubleshoot and maintain. It is recommend to use local variables rather than global variables in the subroutine, because there is almost no address conflict between local variables and other POUs. You can easily transplant the subroutine to other projects.

2) Creation of subroutine

Subroutine is created in the following ways: open the program editor and execute the command "Insert"→"Subroutine" in the "Edit" menu; or click the right mouse button in the program editor window and execute the command "Insert" from the pop-up menu→ "Subroutine". The program editor will automatically generate and open a new subroutine. Right-click the icon of the subroutine or interrupt program in the instruction tree and select "Rename" in the pop-up menu to modify their names. The subroutine can be called to use together with parameters. The parameters are defined in the local variable table of the subroutine. Up to 16 parameters can be transferred. The variable name of the parameter can be up to 23 characters. The symbol "#" in the subroutine variable name represents a local variable, which is automatically added by the programming software. There is no need to input the symbol "#" when inputting local variables manually. It is unavailable to use jump statements to jump into or out of subroutine.

3) Calling of subroutine

You can call subroutine in the main program or interrupt programs. When a part of subroutine is called, all instructions in the subroutine will be executed, and then the pointer will return to the instruction of next subroutine in the program that called it. When inserting a subroutine calling instruction in the ladder program, first open the POU calling to invoke the subroutine in the program editor window, and display the position to insert the subroutine. Use the mouse to double-click to open the "Call subroutine" folder at the bottom of the instruction tree. Press and hold the icon of the subroutine you want to call with the left mouse button, and "drag" it to the desired position in the program editor. Release the left button, the subroutine block is placed at this position. You can also place the rectangular cursor in the position of the program editor window where the subprogram needs to be placed, double-click the subprogram to be called in the instruction tree, and the subprogram block will automatically appear at the position of the cursor.

4) Conditional return of subroutine

In the subroutine, the contact circuit is used to control the CRET (conditional return from the subroutine) instruction. When the contact circuit is turned on, the condition is met and the calling of subroutine is terminated.

5) Timer in the subroutine

When you stop calling the subprogram(subroutine, the ON and OFF state of the bit component of the coil in the subprogram remains unchanged. If the timer in the subroutine is timing when the call is stopped, the 100 ms timer will stop timing, the current value remains unchanged, and the timing will continue when called again. But the 1 ms timer and 10 ms timer will continue to timing until the timed stop time. Their timer bit becomes state 1, and they can work outside the subroutine.

✍ Practice and operation

1. Task preparations

Prepare the components needed for this training task according to Table 4-13.

Table 4-13 Component preparation list

No.	Name	Specification model	Amount
1	Programmable controllers	CPU226	1
2	Four-floor elevator control system unit	YL360-04	1
3	Download cable	PPI cable	1
4	Connection line	YL360 training platform dedicated	Appropriate amount
5	Computer	Self-provided by training room	1

2. Training content

The control panel of four-floor elevator control system is shown in Figure 4-14.

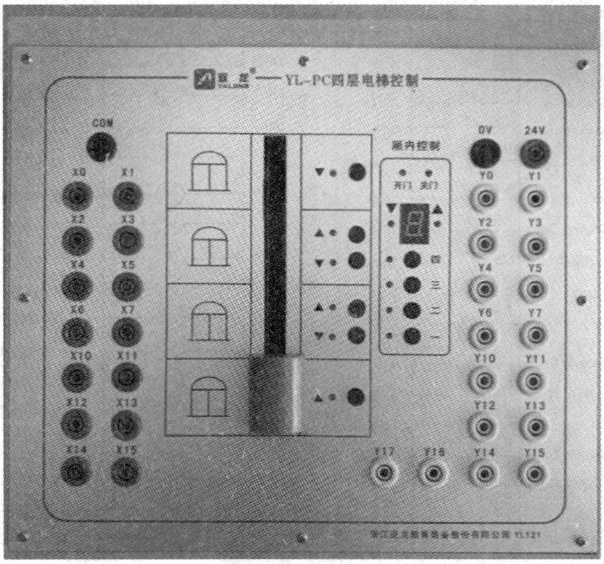

Figure 4-14 Control panel of four-floor elevator control system

The training content of four-floor elevator control system is arranged as follows:

(1) According to the components of the four-floor elevator control system unit, compile the I/O allocation table of the four-floor elevator control system as shown in Table 4-14.

Table 4-14 I/O allocation table of the four-floor elevator control system

Input		Output	
I0.0	Position switch SQ1 on the 1st floor	Q0.0	Upward indication
I0.1	Position switch SQ2 on the 2nd floor	Q0.1	Downward instruction
I0.2	Position switch SQ3 on the 3rd floor	Q0.2	Upward drive
I0.3	Position switch SQ4 on the 4th floor	Q0.3	Down drive
I0.4	Command switch K10 on the 1st floor	Q0.4	Order registration on the 1st floor
I0.5	Command switch K9 on the 2nd floor	Q0.5	Order registration on the 2nd floor
I0.6	Command switch K8 on the 3rd floor	Q0.6	Order registration on the 3rd floor
I0.7	Command switch K7 on the 4th floor	Q0.7	Order registration on the 4th floor
I1.0	Up button K1 on the 1st floor	Q1.0	Up call registration on the 1st floor
I1.1	Up button K3 on the 2nd floor	Q1.1	Up call registration on the 2nd floor
I1.2	Up button K5 on the 3rd floor	Q1.2	Uplift call registration on the 3rd floor
I1.3	Down button K2 on the 2nd floor	Q1.3	Down call registration on the 2nd floor
I1.4	Down button K4 on the 3rd floor	Q1.4	Down call registration on the 3rd Floor
I1.5	Down button K6 on the 4th floor	Q1.5	Down call registration on the 4th floor
		Q1.6	Door opening simulation
		Q1.7	Door closing simulation

(2) Carry out system wiring in accordance with the PLC control program of the four-floor elevator control system (as shown in the Figure 4-15).

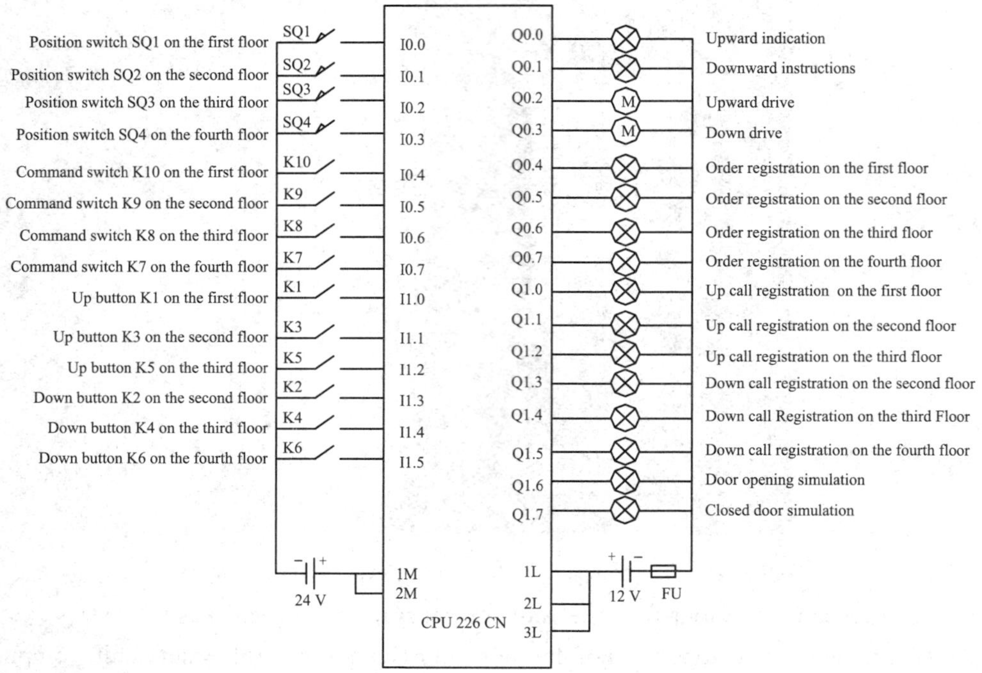

Figure 4-15 PLC control program of the four-floor elevator control system

(3) Compile the PLC control program of the four-floor elevator control system as shown in Figure 4-16.

(4) Download the system program, and debug and run it.

3. Task procedures

(1) First turn the power switch on the PLC host to the OFF state, connect the wires strictly as shown in Figure 4-15. Pay attention to the positive and negative of the 24 V power supply not be short-circuited, otherwise, the PLC contacts will be damaged.

(2) Insert the power cord into the power hole on the surface of the PLC host, and then plug the other end into the 220V power strip.

(3) Turn the power switch of the PLC host to the ON state, and the PLC serial port must be placed in the STOP state. Download the program to the PLC through the computer or programmer. The program of PLC control program of the four-floor elevator control system is shown in Figure 4-15. The PLC serial port is placed in the RUN state.

(4) Operate in accordance with the principle of practical training and observe the phenomena in practical training.

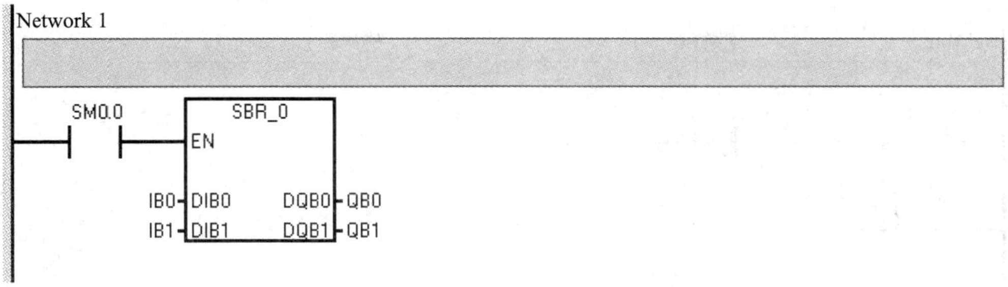

(a) Main program diagram

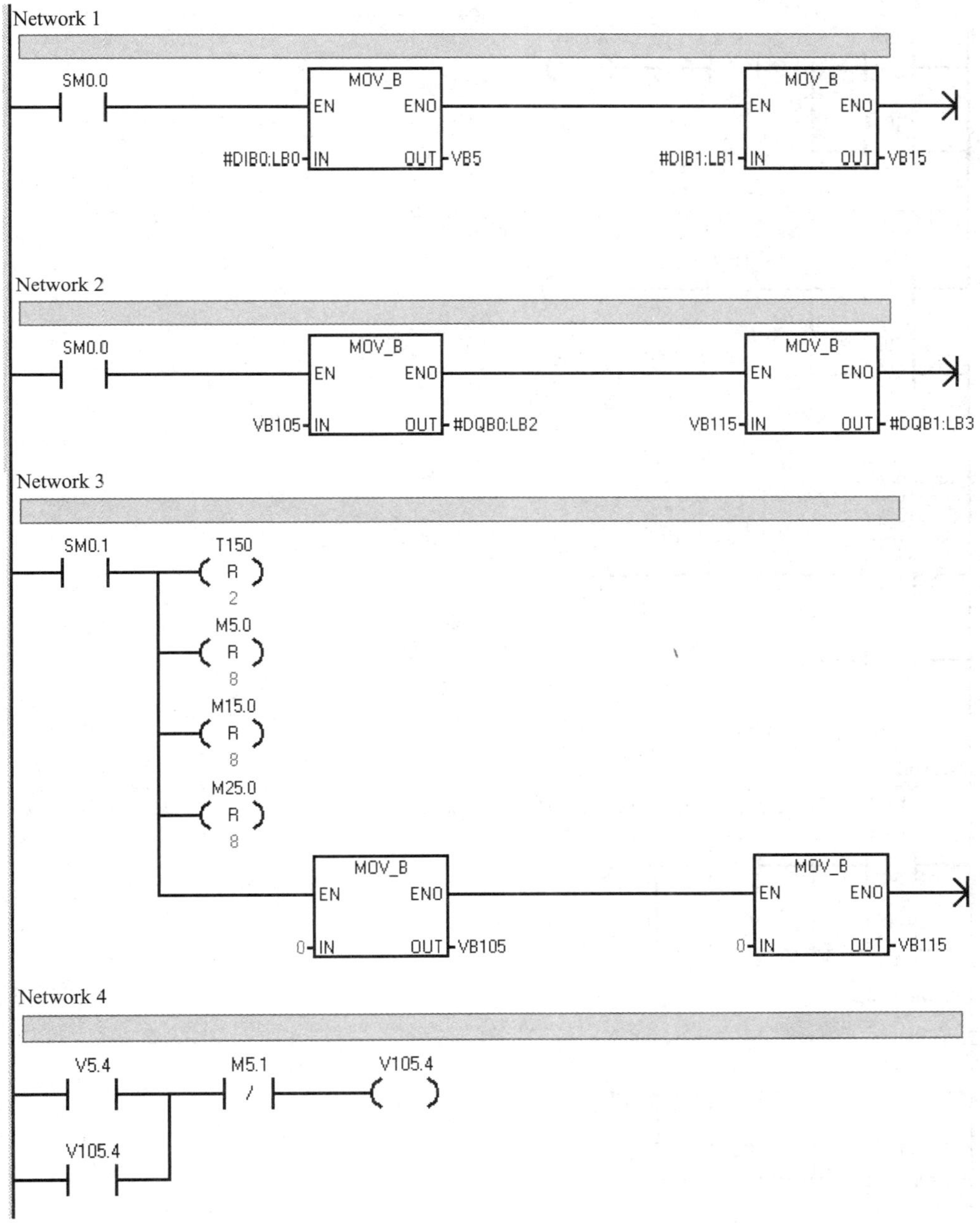

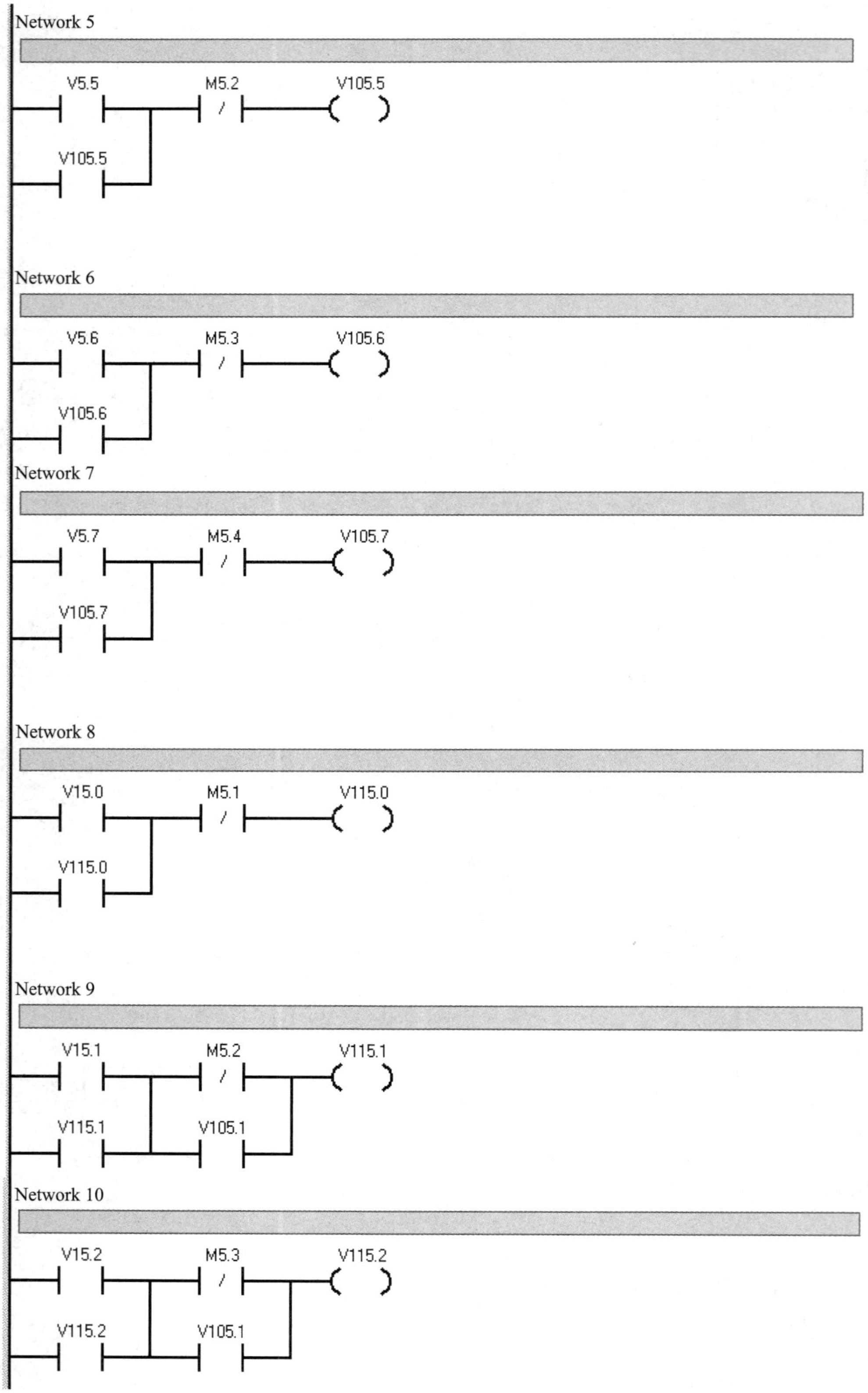
Network 5
V5.5
M5.2
V105.5
V105.5
Network 6
V5.6
M5.3
V105.6
V105.6
Network 7
V5.7
M5.4
V105.7
V105.7
Network 8
V15.0
M5.1
V115.0
V115.0
Network 9
V15.1
M5.2
V115.1
V115.1
V105.1
Network 10
V15.2
M5.3
V115.2
V115.2
V105.1

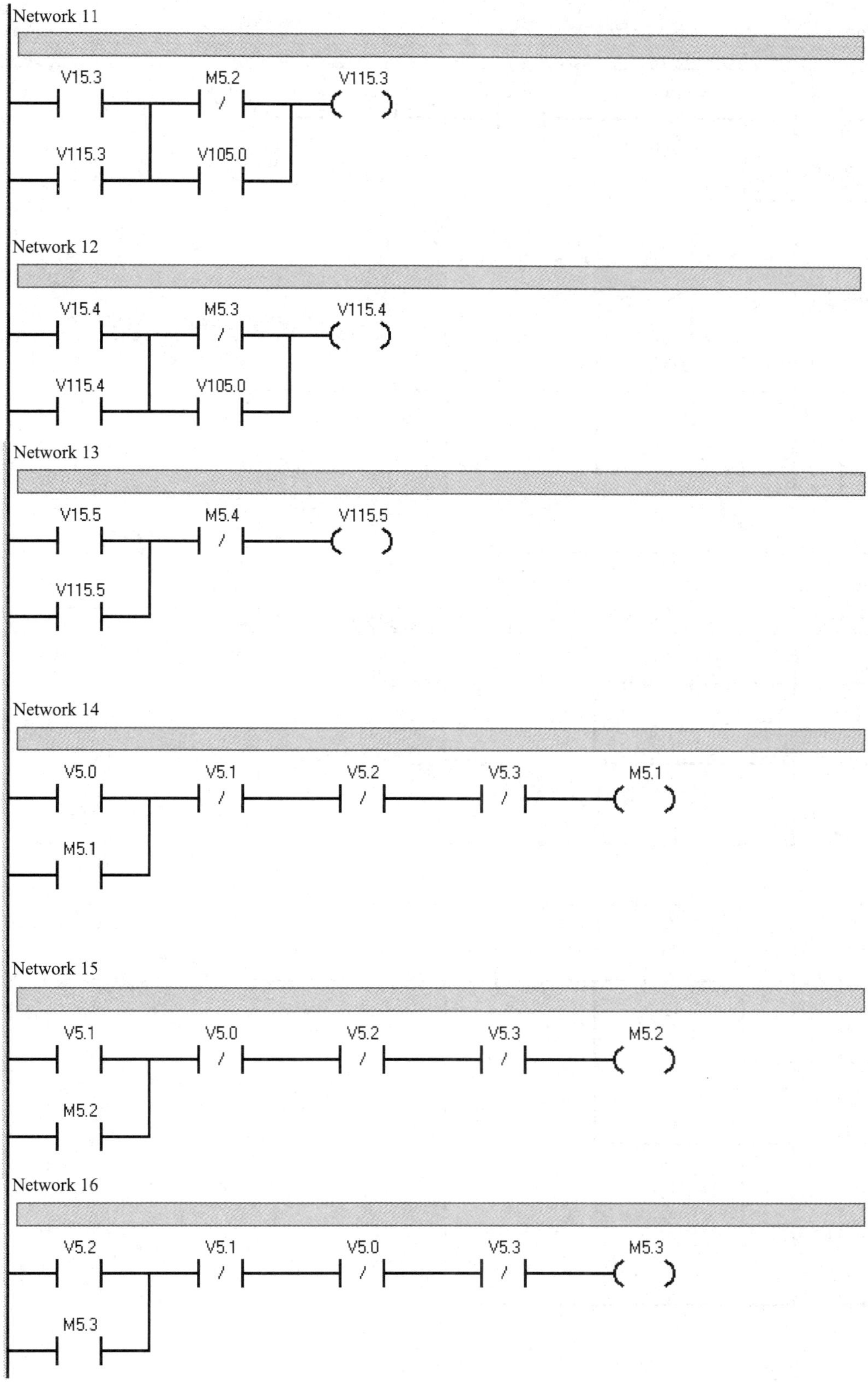
Network 11
V15.3
M5.2
V115.3
V115.3
V105.0
Network 12
V15.4
M5.3
V115.4
V115.4
V105.0
Network 13
V15.5
M5.4
V115.5
V115.5
Network 14
V5.0
V5.1
V5.2
V5.3
M5.1
M5.1
Network 15
V5.1
V5.0
V5.2
V5.3
M5.2
M5.2
Network 16
V5.2
V5.1
V5.0
V5.3
M5.3
M5.3

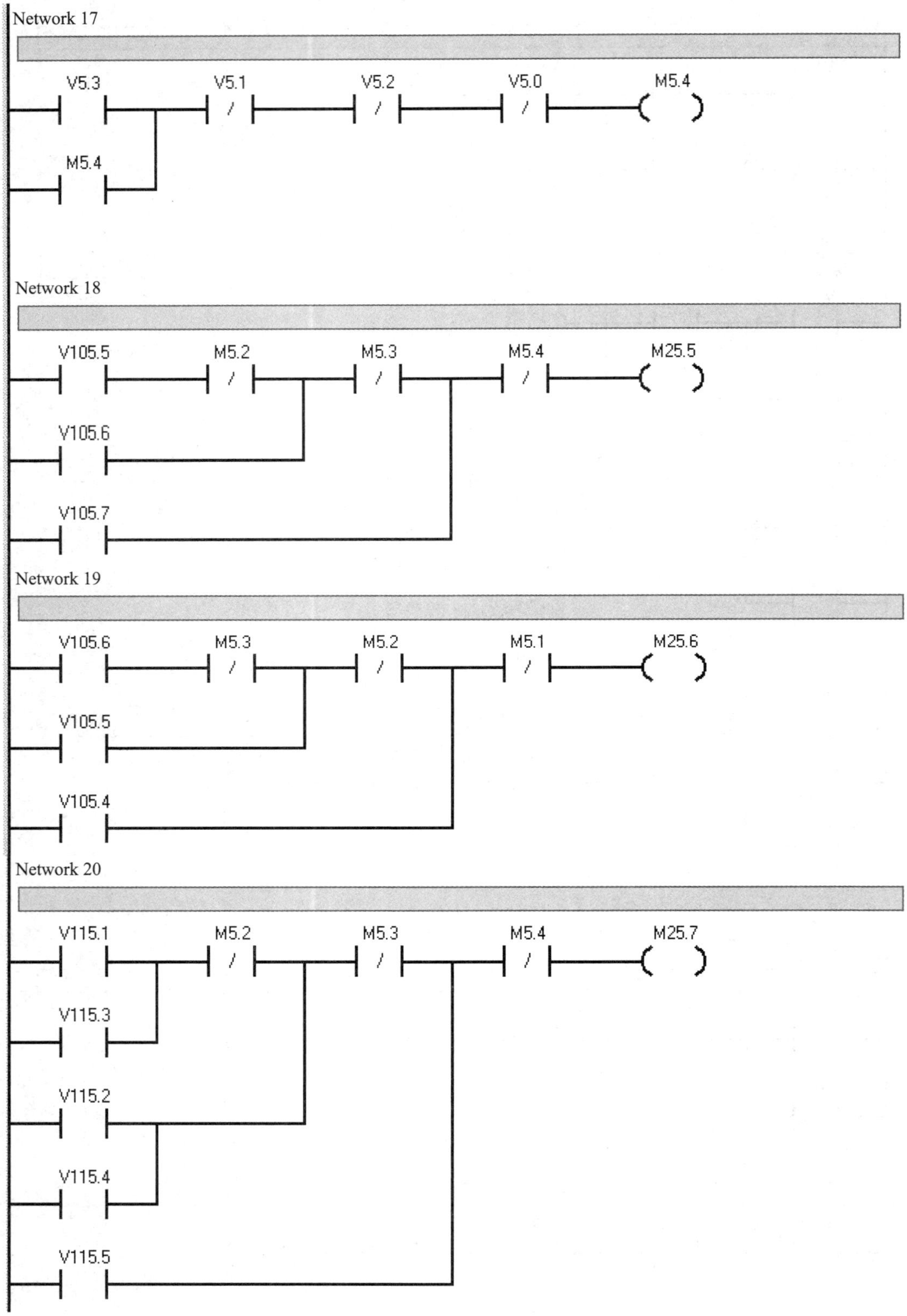
Network 17
V5.3
V5.1
V5.2
V5.0
M5.4
M5.4
Network 18
V105.5
M5.2
M5.3
M5.4
M25.5
V105.6
V105.7
Network 19
V105.6
M5.3
M5.2
M5.1
M25.6
V105.5
V105.4
Network 20
V115.1
M5.2
M5.3
M5.4
M25.7
V115.3
V115.2
V115.4
V115.5

Network 21

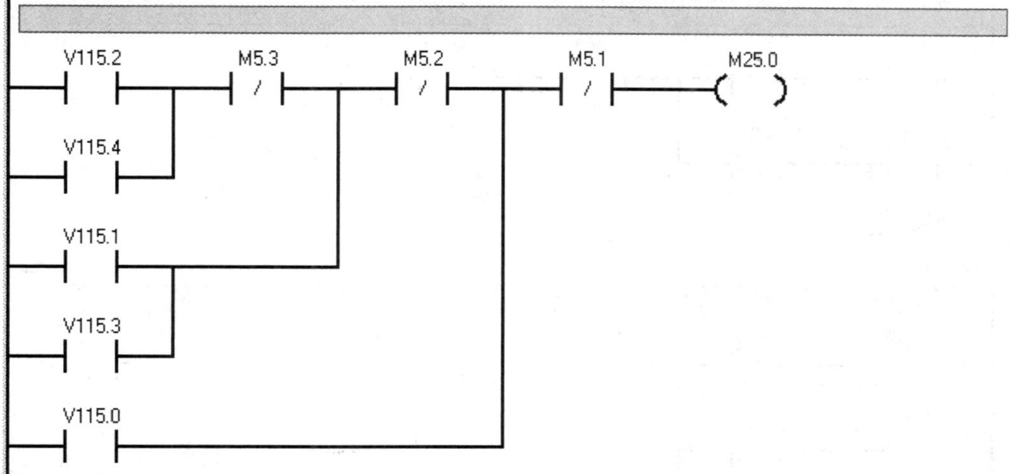

Network 22

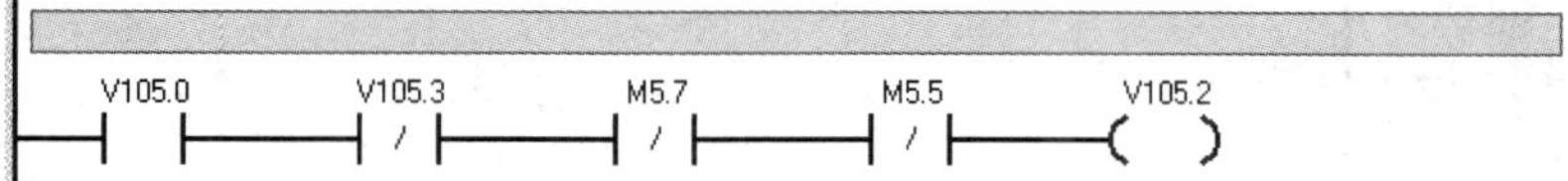

Network 23

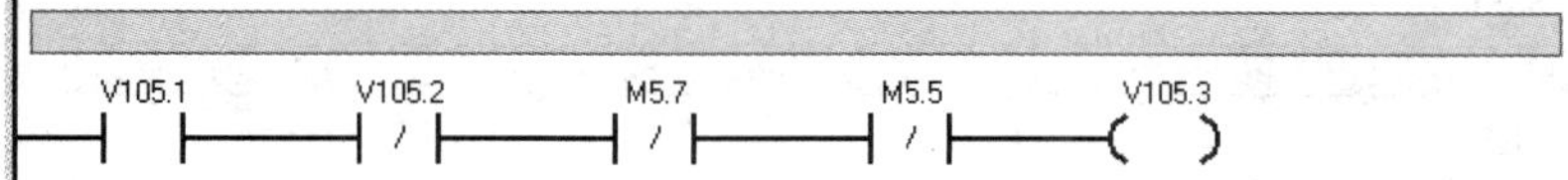

Network 24

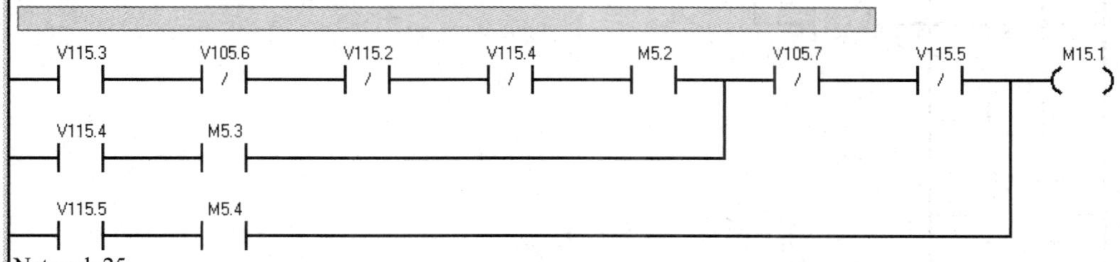

Network 25

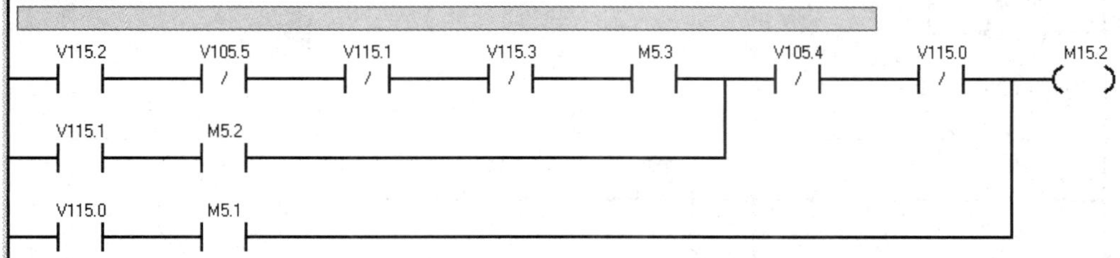

Network 26

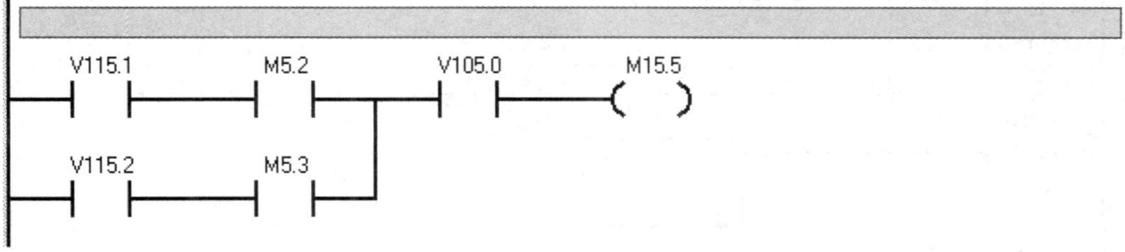

Network 27

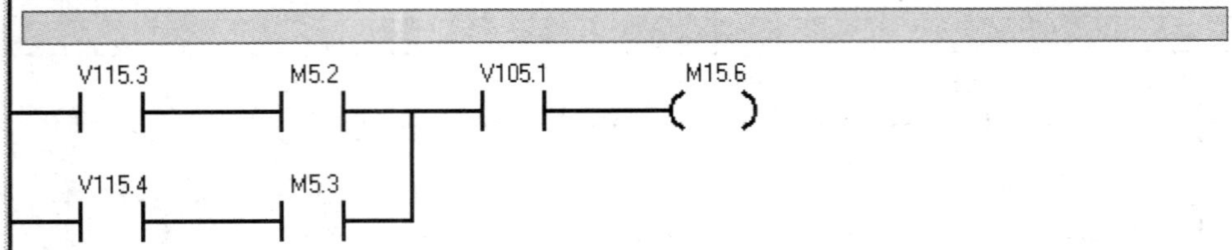

Network 28

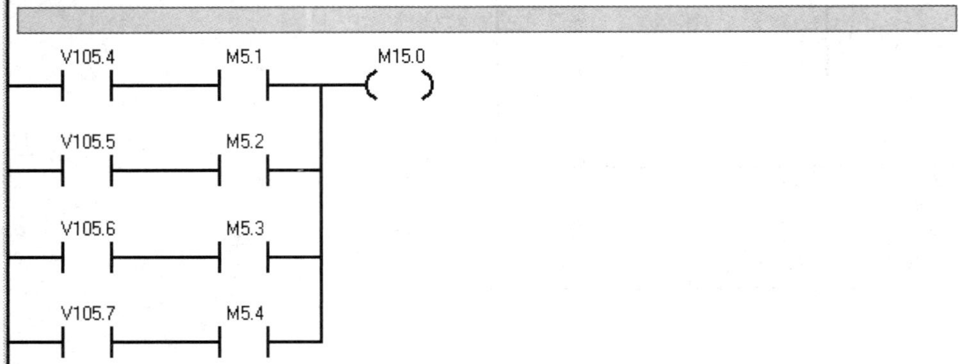

Network 29

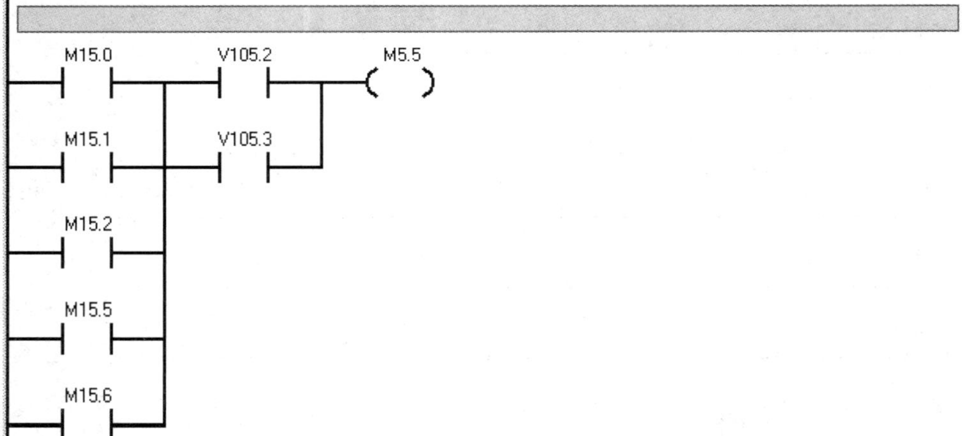

Network 30

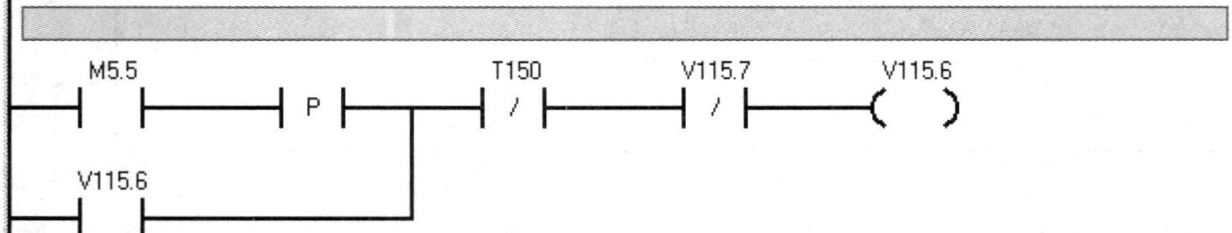

Network 31

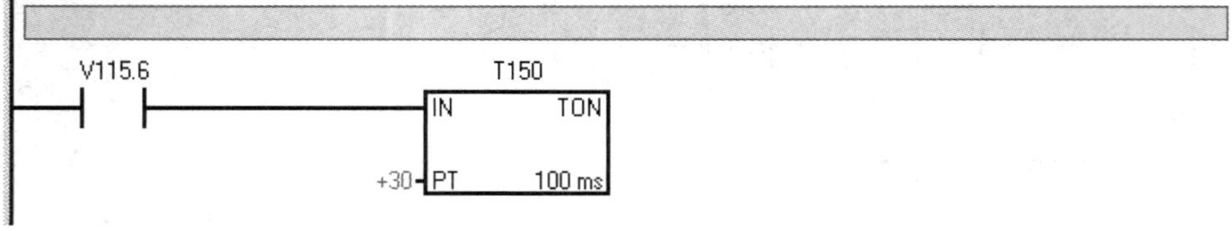

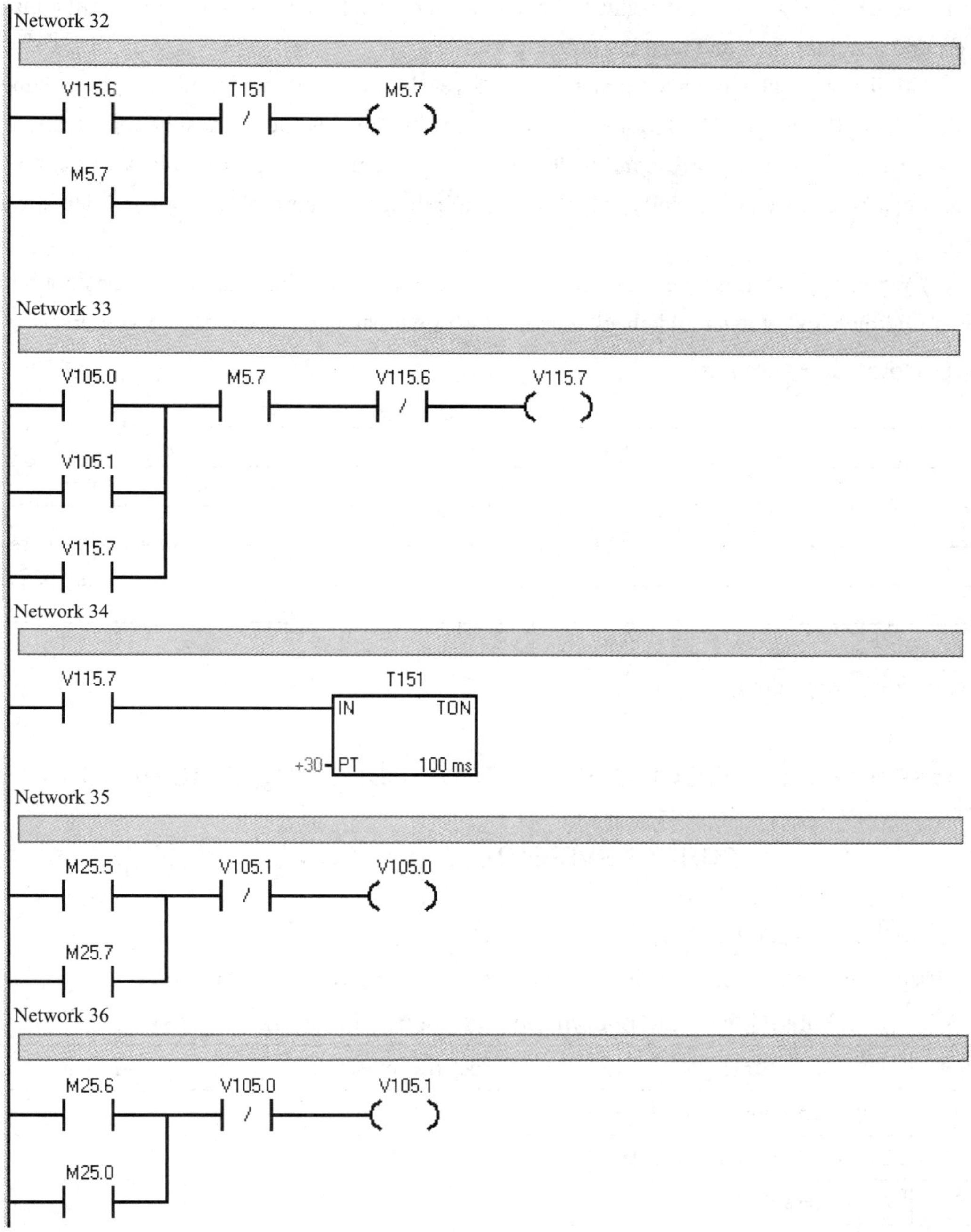

(b) Subroutine diagram

Figure 4-16 PLC control program of the four-floor elevator control system

4. Precautions

(1) Do not touch the live parts, strictly follow the operating sequence of "wiring first, then power on; first connect the control circuit, then connect the power supply; disconnect the power first, then disconnect the wires".

(2) When wiring, you must connect the load terminal first, then the power terminal; connect the ground terminal first, and then the power cord.

(3) If abnormal phenomena (such as noise, fever, burnt smell) are found, the power supply should be cut off immediately; the scene must be kept, and the instructor must be reported.

(4) The wiring must correspond to the electrical schematic diagram, especially the power wiring must be connected correctly, which will cause the power unit to burn out and damage the unit.

(5) When downloading the program, if you encounter failure information such as not being able to find the CPU, you should check whether the download cable is connected properly.

5. Notes and summary

____

____

____

____

____

____

【Extended Practices】

Exercise 1　Programming and debugging of tower light control system

1. Task preparations

Prepare the components needed for this training task according to Table 4-15.

Table 4-15　List of components needed in the training task

No.	Name	Specification model	Amount
1	Programmable controllers	CPU226	1
2	Tower light control system unit	YL360-05	1
3	Download cable	PPI cable	1
4	Connection line	YL360 training platform dedicated	Appropriate amount
5	Computer	Self-provided in the training room	1

2. Training content

The training content of the tower light control system is arranged as follows.

(1) According to the components of the tower light control system, compile the I/O allocation table of the tower light control system as shown in Table 4-16.

(2) Carry out the system wiring according to the PLC control principle diagram of the tower

light control system shown in Figure 4-17.

(3) Compile the PLC control program of the tower light control system.

(4) Download the control program, and debug and run it.

Table 4-16 I/O allocation table of the tower light control system

Input		Output	
Q0.0	Lantern L1	Q1.1	Seven-segment decoding tube A
Q0.1	Lantern L2	Q1.2	Seven-segment decoding tube B
Q0.2	Lantern L3	Q1.3	Seven-segment decoding tube C
Q0.3	Lantern L4	Q1.4	Seven-segment decoding tube D
Q0.4	Lantern L5	Q1.5	Seven-segment decoding tube E
Q0.5	Lantern L6	Q1.6	Seven-segment decoding tube F
Q0.6	Lantern L7	Q1.7	Seven-segment decoding tube G
Q0.7	Lantern L8		
Q1.0	Lantern L9		

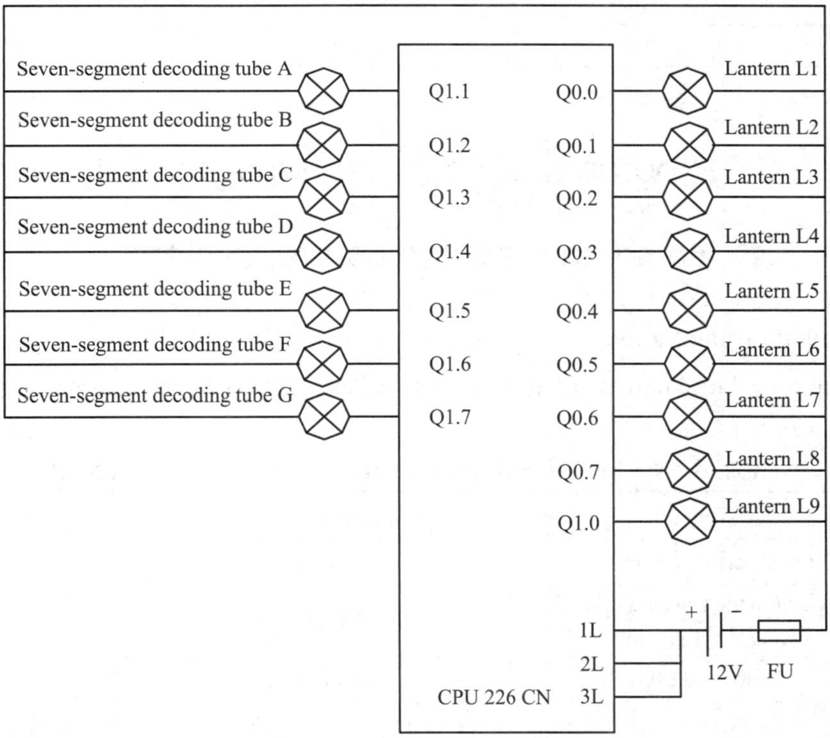

Figure 4-17 The PLC control principle diagram of the tower light control system

3. Task procedures

Task procedures are as follows.

(1) First, set the power switch on the PLC host to OFF state, and do wiring strictly

according to Figure 4-17. Note that the positive and negative of the 24V power supply should not be short-circuited, otherwise, the PLC contacts will be damaged.

(2) Insert the power cord into the power hole on the surface of the PLC host, and then plug the other end into the 220 V power strip.

(3) Turn the power switch of the PLC host to ON state. The PLC serial port must be placed in the STOP state, and then download the programmed program to the PLC through the computer or programmer. After the download is complete, place the PLC serial port in RUN state.

(4) After the PLC is running, observe the light status of the controlled object.

4. Precautions

Refer to Tasks 1-3 in this module.

5. Notes and summary

____

____

____

____

____

____

Exercise 2　Programming and debugging of fully automatic washing machine control system

1. Training preparations

According to the requirements of this training tasks, prepare the required components as shown in Table 4-17.

Table 4-17　List of components needed in the training task

No.	Name	Specification model	Amount
1	Programmable controllers	CPU226	1
2	Fully automatic washing machine control system unit	YL-360-05	1
3	Download cable	PPI cable	1
4	Connection line	YL-360 training platform dedicated	Appropriate amount
5	Computer	Self-provided training room	1

2. Training content

The training content of fully automatic washing machine control system is arranged as follows.

(1) According to the components of the fully automatic washing machine control system unit, compile the I/O allocation table of the fully automatic washing machine control system as shown in Table 4-18.

(2) Carry out the system wiring according to the PLC control principle diagram of fully automatic washing machine control system as in Figure 4-18.

(3) Compile the PLC control program of the fully automatic system.

(4) Download the control program, and debug and run it.

Table 4-18 I/O allocation table of fully automatic washing machine control system

Input		Output	
I0.0	Run button	Q0.0	Water inlet indicator
I0.1	Stop button	Q0.1	Drain indicator
I0.2	Upper limit button	Q0.2	Straight polarity indicator
I0.3	Lower limit button	Q0.3	Reversed polarity indicator
I0.4	Manual drain button	Q0.4	Dry bucket indicator
		Q0.5	Buzzer indicator

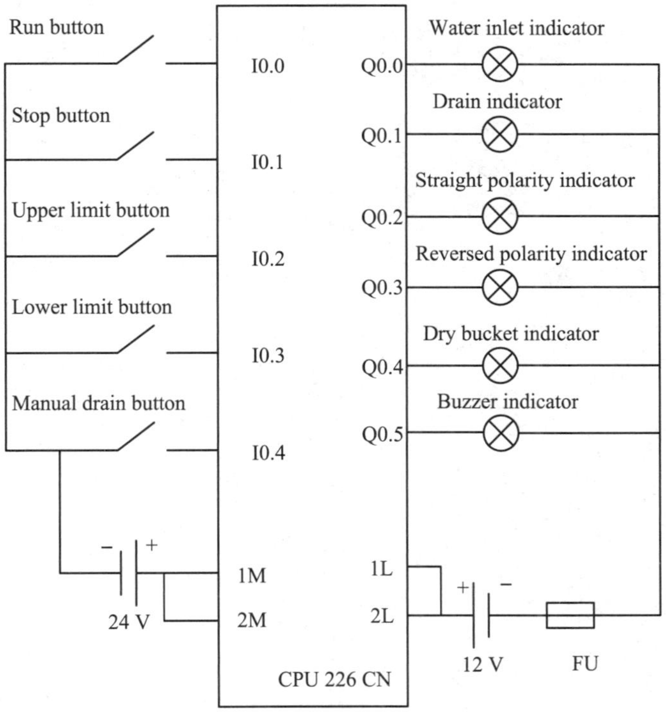

Figure 4-18 PLC control principle diagram of fully automatic washing machine control system

3. Task procedures

Task procedures are as follows.

(1) First make the power switch on the PLC host in OFF state, and do wiring strictly as

shown in Figure 4-18. Note that the positive and negative of the 24 V power supply should not be short-circuited, otherwise, the PLC contacts will be damaged.

(2) Insert the power cord into the power hole on the surface of the PLC host, and then plug the other end into the 220 V power strip.

(3) Make the power switch on the PLC host in ON state, and the PLC serial port must be placed in the STOP state, and then download the program to the PLC through the computer or programmer, and then set the PLC serial port in RUN state after the download is finished.

(4) Operate according to the principle of training.

4. Precautions

Refer to Tasks 1-3 in this module.

5. Notes and summary

____

____

____

Module 5 Programming and debugging of YL-335B automatic production line

【General Introduction】

This module carries out programming and debugging training of YL-335B automatic production line. YL-335B automatic production line training equipment (YL-335B equipment) is composed of 5 units: feeding unit, processing unit, assembly unit, conveying unit and sorting unit installed on the aluminum alloy rail-type training platform. Their appearance is shown in Figure 5-1. Among them, each work unit can form an independent mechatronics system by itself. Each unit is basically pneumatic actuator, but the overall movement of the manipulator device of the conveying unit is driven by a stepping motor in precise positioning control. The driving system has the characteristics of long stroke and multiple positioning points, which is a typical example of one-dimensional positioning control system. The conveyor belt of the sorting unit adopts an AC transmission device (a three-phase asynchronous motor driven by a general frequency converter).

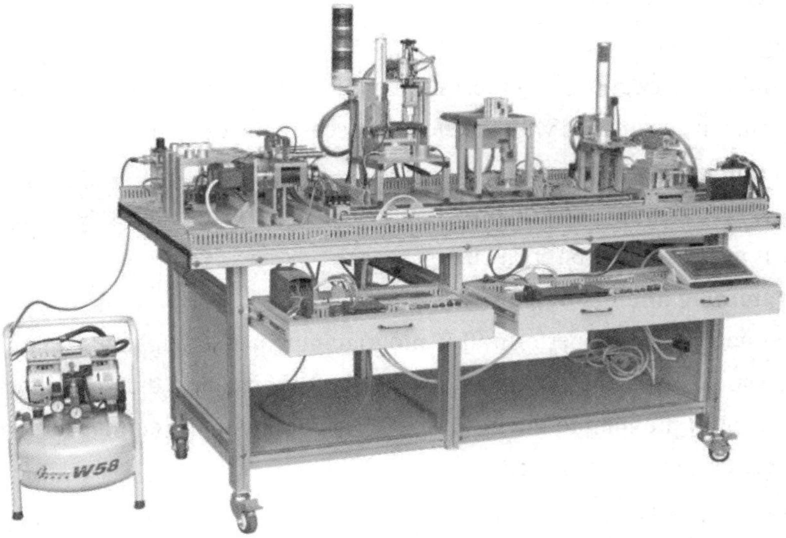

Figure 5-1 Appearance of YL-335B equipment

There are three tasks in this module. The tasks are as follows:

(1) Programming and debugging of the feeding unit control system.

(2) Programming and debugging of the processing unit control system.

(3) Programming and debugging of the assembly unit control system.

Through the training of the above three tasks, it is planned to improve the application ability of senior electricians, such as programming and debugging of automation controllers, and to meet the skill requirements of senior electricians.

【Training Objectives】

1. Knowledge objectives

(1) Master the working process of each part of YL-335B equipment.

(2) Master the use of pneumatic components in YL-335B equipment.

(3) Master the electrical schematic diagram of YL-335B equipment.

(4) Master the control program structure and programming ideas of the feeding unit.

(5) Master the control program structure and programming ideas of the processing unit.

(6) Master the control program structure and programming ideas of the assembly unit.

2. Skill objectives

(1) Able to process the I/O wiring of the feeding unit control system.

(2) Able to process the PLC programming and debugging of the feeding unit control system.

(3) Able to process perform I/O wiring of the processing unit control system.

(4) Able to process the PLC programming and debugging of the processing unit control system.

(5) Able to process perform I/O wiring of the assembly unit control system.

(6) Able to process the PLC programming and debugging of the assembly unit control system.

3. Operating standards

1) Wiring

Adhere to the electrical wiring process standards, perform the installation layout and wiring of the components in accordance with the electrical schematic diagram of the task.

2) Inspection and investigation

Perform a test before powering on the wired electrical circuit, focusing on checking whether there is a short circuit or open circuit.

3) Programming and debugging

Power on the system and perform PLC programming and debugging. Before compiling the program, you should draw the program flow chart to ensure that the compiling idea is clear. Download the control program to the PLC, operate according to the equipment working process, and complete the task of system operation.

4. Safety standards

In order to ensure the personal safety and training safety during the training, the training members must strictly abide by the following safety standards.

(1) The trainer should ensure that the device has reliable protective grounding, leakage protection, overload protection, insulated table tops, insulated carpets and safety patch cords.

(2) During the training, the human body should not be in contact with bare live objects.

(3) Wiring and disconnecting must be done with the power supply cut off.

(4) PLC wiring must be carried out strictly in accordance with the electrical control schematic diagram, and the positive and negative power supply cannot be connected reversely.

(5) Before powering on the system, it is necessary to check the power supply wiring.

5. Quality standards

(1) Adjust the pneumatic part when the system is powered on, check whether the air circuit is normal, and ensure that the air pressure is between 0.4–0.5MPa.

(2) Check whether the installation position of the magnetic switch is in place to ensure that the magnetic switch is working normally.

(3) Ensure that the I/O wiring is completely consistent with the electrical schematic diagram.

(4) Check whether the photoelectric sensor (switch) is installed properly, and adjust the test detection sensitivity knob to ensure the reliability of the test.

【Training Requirements】

1. Component preparation

List of components to prepare is as Table 5-1.

Table 5-1 Component preparation list

No.	Name	Unit	Amount
1	YL-335B automatic production line training equipment	Set	1
2	Computer	Set	1
3	Siemens S7-200 programming cable	Piece	1
4	Connection line	Piece	20
5	Digital multimeter	Piece	

2. Knowledge identification point

(1) The use of pneumatic components in YL-335B equipment.

(2) The electrical schematic diagram of YL-335B equipment.

(3) The control program structure and programming ideas of the feeding unit.

(4) The control program structure and programming ideas of the processing unit.

(5) The control program structure and programming ideas of the assembly unit.

3. Skills assessment point

(1) Perform I/O wiring of the feeding unit control system.

(2) Perform PLC programming and debugging of the feeding unit control system.

(3) Perform I/O wiring of the processing unit control system.

(4) Perform PLC programming and debugging of the processing unit control system.

(5) Perform I/O wiring of the assembly unit control system.

(6) Perform PLC programming and debugging of the assembly unit control system.

【Work Tasks】

Task 1 Programming and debugging of feeding unit control system

✍ Relevant knowledge

1. Basic functions of automatic production line

The top view of YL-335B equipment is shown in the Figure 5-2.

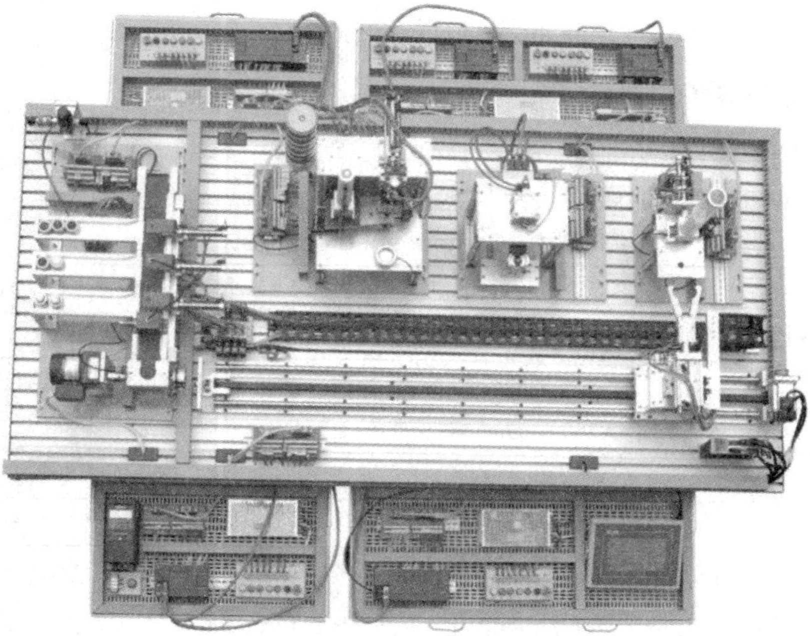

Figure 5-2 Top view of YL-335B equipment

1) The basic function of the feeding unit

The feeding unit undertakes the first task in YL-335B equipment system, it works in supplying raw materials to other units in the system. The specific function of the feeding unit is to automatically push the workpiece (raw material) in the silo to be processed onto the material table, prepared for the manipulator of the conveying unit to grab the raw material and convey it to other units. Figure 5-3 shows the overall view of the feeding unit.

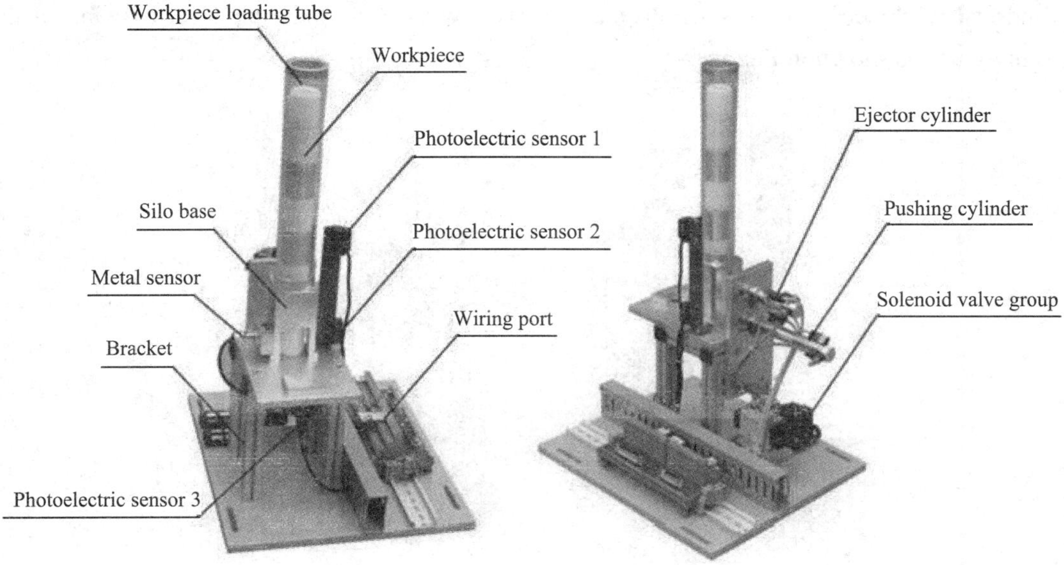

(a) Front view (b) Side view

Figure 5-3 Overall view of the feeding unit

2) The basic function of the processing unit

The processing unit sends the workpiece on the material table (by the gripping manipulator of the conveying unit) to backside of the punching body, completing a punching processing action, and then sends it back to the material table, waiting to be taken out by the gripping manipulator of the conveying unit . Figure 5-4 shows the overall view of the processing unit.

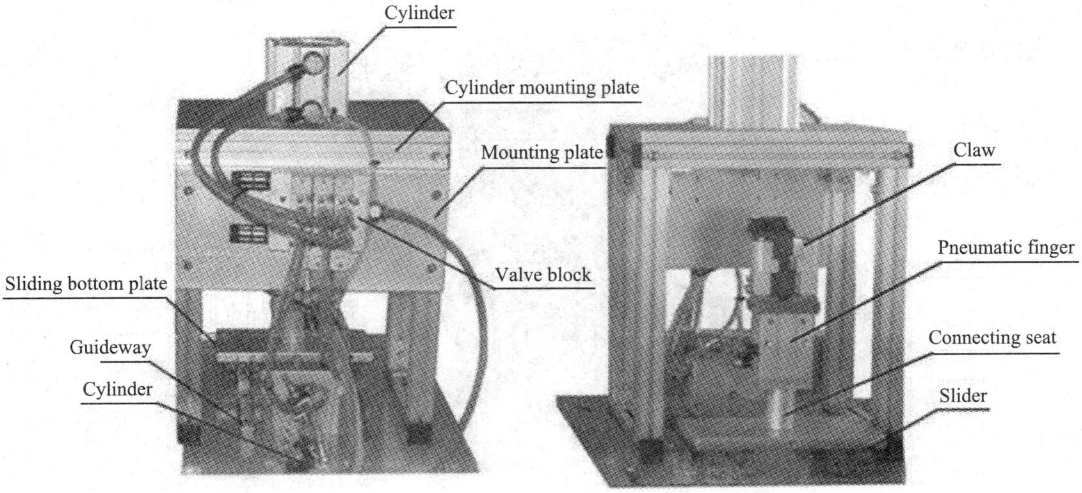

(a) Back view (b) Front view

Figure 5-4 Overall view of the processing unit

3) The basic function of the assembly unit

The assembly unit completes the assembly process: embedding small black or white cylindrical workpieces in the silo into the processed workpieces. The physical drawing of the assembly unit is shown in Figure 5-5.

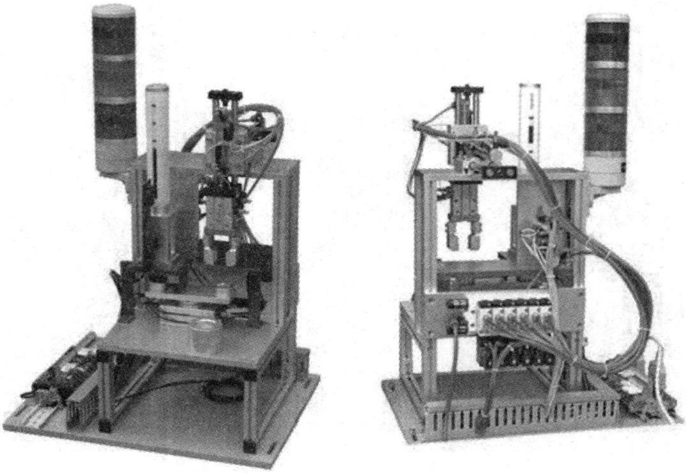

(a) Front view　　(b) Back view

Figure 5-5　Physical drawing of assembly unit

4) The basic function of the sorting unit

The sorting unit sorts the processed and assembled workpieces from the previous unit, ready to make the workpieces of different colors in different troughs diverted. Figure 5-6 shows the complete view of the sorting unit.

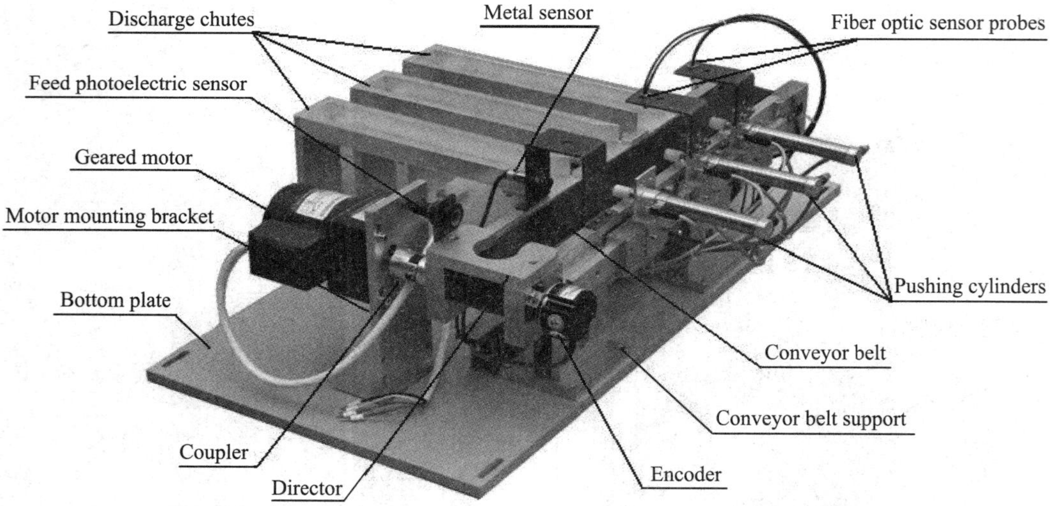

Figure 5-6　Complete view of the sorting unit

5) The basic function of the conveying unit

The conveying unit drives the gripping manipulator to the material table of the specified unit supported by linear motion conveying mechanism and accurate positing system. The unit grabs the workpiece on the material table, transports it to specified location and then puts it down, realizing the conveying function. The linear motion conveying mechanism is driven by a servo motor. The appearance of the conveying unit is shown in Figure 5-7.

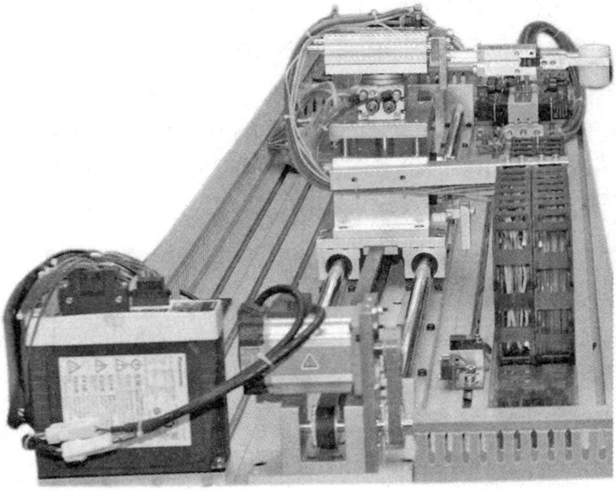

Figure 5-7 Appearance of conveying unit

2. Electric control of YL-335B equipment

1) Structural characteristics of YL-335B equipment working unit

The structural feature of each working unit in YL-335B equipment is that the mechanical devices and electrical control parts are relative independent. The mechanical device of each work unit is integrally installed on the bottom plate, and the PLC device controlling the production process of the work unit is installed on the drawer plates on both sides of the workbench. Therefore, the information exchange between the mechanical device and the PLC device is a key issue. YL-335B training equipment connects the lead cable of the solenoid valves and sensors on the mechanical device to the wiring ports on the device side. The I/O lead wire of the PLC is connected to the wiring port on the PLC side. The two wiring ports are interconnected by a multi-core signal cable. Figure 5-8 and Figure 5-9 are the wiring port on the device side and the wiring port on the PLC side respectively.

Figure 5-8 Wiring port on the device side

Figure 5-9 Wiring port on the PLC side

The wiring terminal of the wiring port on the device side adopts a three-layer terminal structure. The upper terminal is used to connect the +24V terminal of the DC24V power supply. The bottom terminal is used to connect the 0V terminal of the DC24V power supply. The middle terminal is used to connect each signal line.

The wiring terminal of the wiring port on the PLC side adopts a two-layer terminal structure. The upper terminal is used to connect each signal line, and its terminal number corresponds to the wiring terminal of the wiring port on the device side. The bottom terminal is used to connect the +24V terminal and the 0V terminal of the DC24V power supply.

The connection port on the device side and the connection port on the PLC side are connected by a dedicated cable. The 25-pin connector cable is connected to the input signal line of the PLC, and the 15-pin connector cable is connected to the output signal line of the PLC.

2) The control system of YL-335B equipment

Each working unit for the control system of YL-335B equipment can form an independent system by itself, and they can also form a distributed control system through network interconnection.

(1) When the work unit forms an independent system by itself, the button indicator module of the work unit displays the main signal when the equipment starts to operation and the status display signal when the equipment is in operation. The button indicator module box is shown in Figure 5-10. The terminal pins of the indicator lights and buttons on the control unit all led to the terminal block.

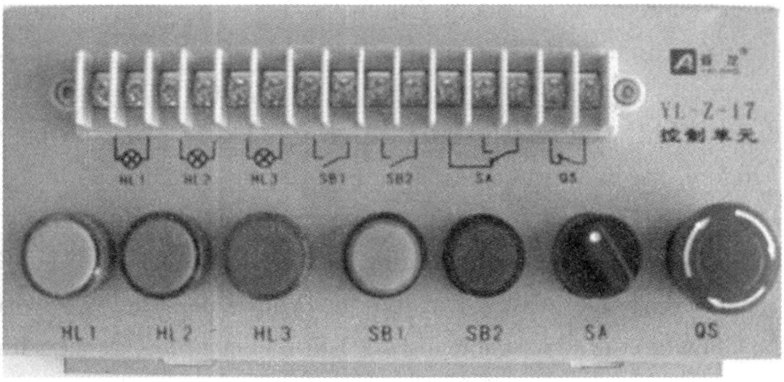

Figure 5-10　Button indicator of control (system) unit

The components on the button indicator module include:

① Indicator light (24V DC): one each for yellow (HL1), green (HL2), and red (HL3).

② Master device: one green normally open button SB1, one red normally open button SB2, selector switch SA (with a pair of change-over contacts), emergency stop button QS (one normally closed contact).

(2) When the working units form a distributed control system through network interconnection, for the equipment using Siemens SMART 200 PLC, the standard configuration of YL-335B equipment adopts the communication mode of TCP/IP protocol. The factory control scheme of the equipment is shown in Figure 5-11.

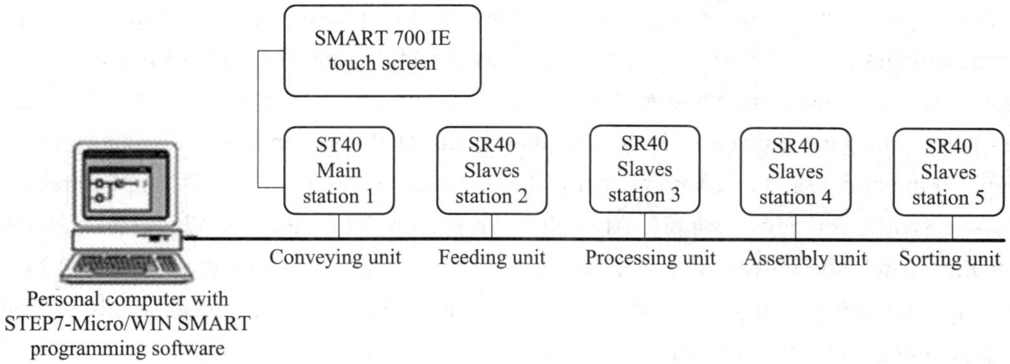

Figure 5-11 Factory control scheme of YL-335B equipment

The PLC configuration of each workstation is as follows:

① Conveying unit: it is SMART200 ST40 DC/DC/DC main unit, with a total of 24 points of input and 16 points of transistor output.

② Feeding unit: it is SMART200 SR40 AC/DC/RLY main unit, with a total of 24 points of input and 16 points of relay output.

③ Processing unit: it is SMART200 SR40 AC/DC/RLY main unit, with a total of 24 points of input and 16 points of relay output.

④ Assembly unit: it is SMART200 SR40 AC/DC/RLY main unit, with a total of 24 points of inputs and 16 points of relay output.

⑤ Sorting unit: it is SMART200 SR40 AC/DC/RLY main unit, with a total of 18 points of input and 16 points of relay output.

(3) Human-machine interface. The main signal (reset, start, stop, etc.) of the system operation is given through the touch screen of human-machine interface. At the same time, the human-machine interface also displays various status information of system operation.

The human-machine interface is a two-way communication bridge between the operator and the machine equipment. Using the human-machine interface can clearly indicate and inform the current status of the equipment to the operator, making the operation simple, vivid and accurate, even a novice can easily operate the entire equipment under rigid instructions. The use of human-machine interface can also standardize and simplify the wiring of the machine. It can also reduce the number of I/O points required by the PLC controller (reduce the production cost). At the same time, due to the miniaturization and high performance of the panel control, the added value of the whole set of equipment is relatively high.

YL-335B equipment adopts Siemens SMART 700IE touch screen as its human-machine interface. SMART 700IE is a high-performance embedded integrated industrial computer with embedded low-power CPU as the core (frequency 400 MHz). The product design uses a high-resolution 64 K true-color widescreen display. Its LED backlight reduces equipment power consumption and prolongs equipment service life. It supports Ethernet and serial communication functions, and the maximum communication rate of 187.5 kb/s.

(4) Power supply. The external power supply is three-phase five-wire. Figure 5-12 is the schematic diagram of the primary circuit of the power supply unit. Figure 5-13 is the installation diagram of the power distribution box. The main power switch selects DZ47LE-32/C32 three-phase four-wire leakage switch. The main loads of the system are individually powered through automatic switches. Among them, the inverter power uses DZ47C16/3P three-phase automatic switch for power supply. At each work station, PLC uses DZ47C5/1P single-phase automatic switch for power supply. In addition, the system is equipped with 4 DC 24V/6A switching stabilized power supply, which is respectively used as the DC power supply of feeding unit, processing unit, assembly unit, sorting unit, and conveying unit.

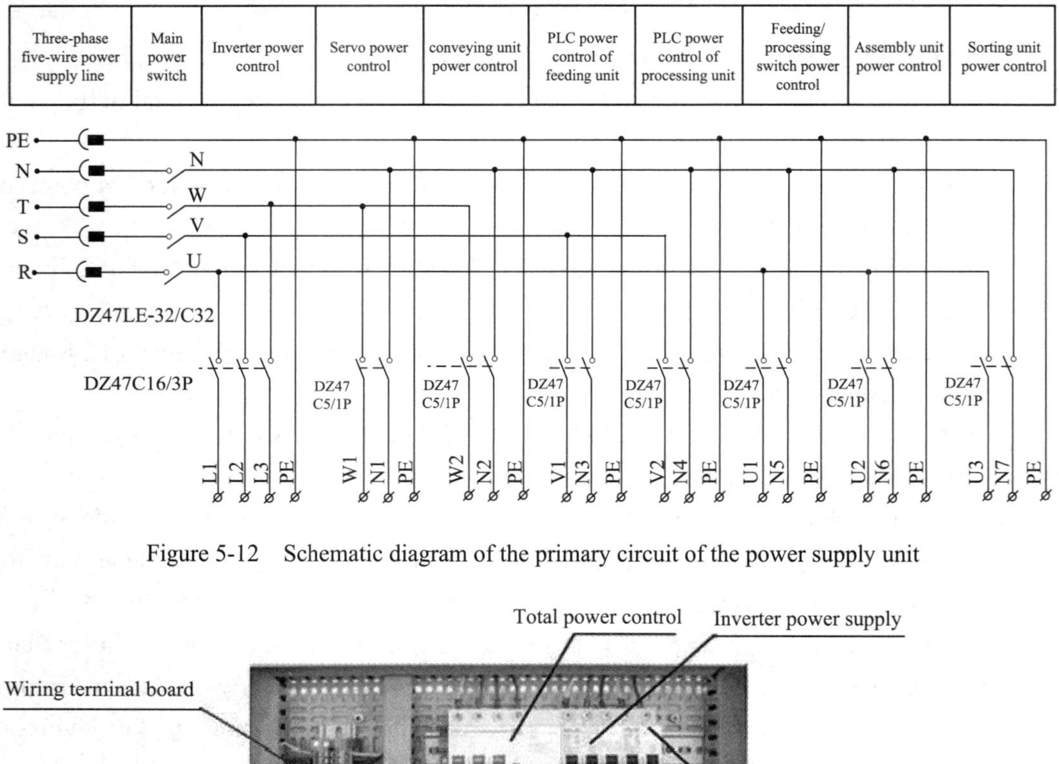

Figure 5-12　Schematic diagram of the primary circuit of the power supply unit

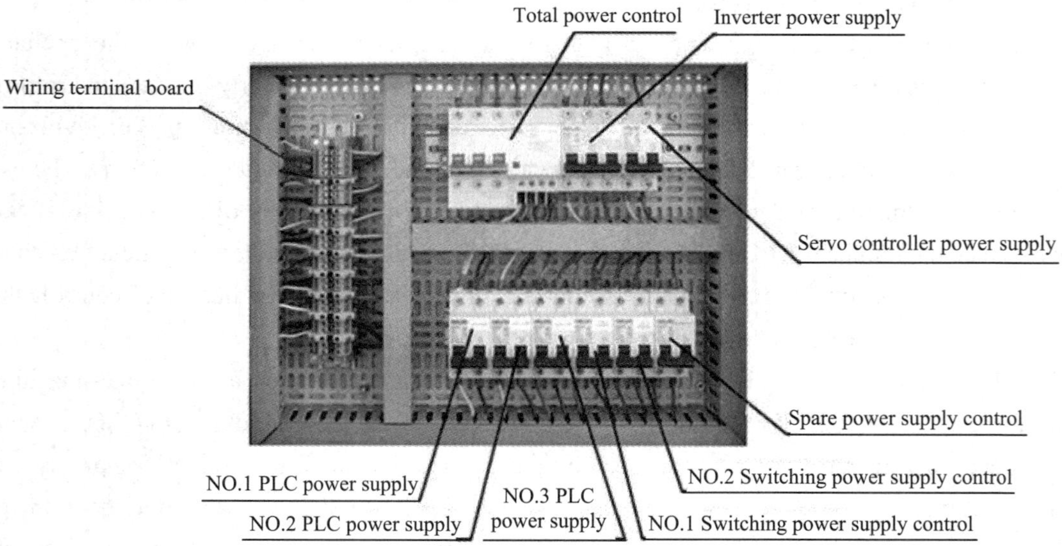

Figure 5-13　Installation diagram of the power distribution box

(5) Air source processing device. The air source processing components and their schematic diagram of YL-335B equipment pneumatic circuit are shown in Figure 5-14. The air source processing components are the basic components of the pneumatic control system. Their function is to remove the impurities and condensate contained in the compressed air, adjust and maintain a constant working pressure. When in use, you should always check the condensed water level in the filter. Before it exceeds the highest mark, it must be discharged to avoid being re-inhaled. A quick switch installed at the air path entrance of the air source processing components is used to open/close the air source. When the air switch is pulled out to the left, the air circuit is open. When the air switch is pulled out to the right, the air circuit is closed. The input air source of the air source processing components comes from an air compressor, and the pressure provided by the gas compressor is 0.6−1.0 MPa, and the output pressure is adjustable in the range of 0−0.8 MPa. The output compressed air is delivered to each working unit through quick three-way joint and air tube.

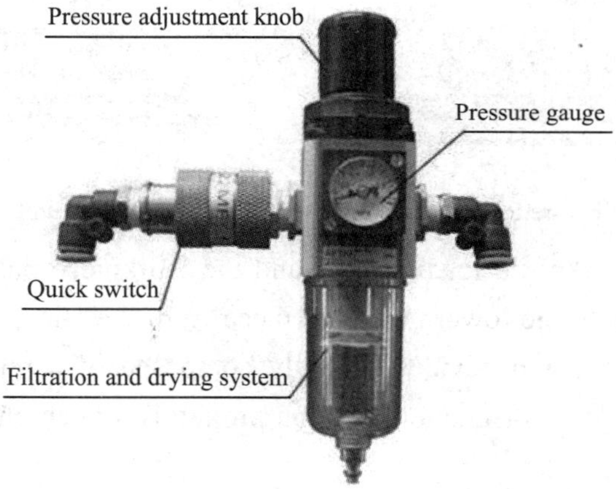

(a) Physical picture of air source processing components

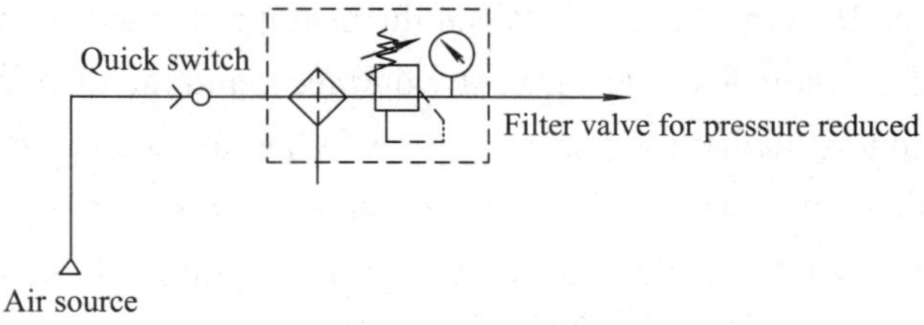

(b) Pneumatic circuit schematic diagram

Figure 5-14 Schematic diagram of air source processing components pneumatic circuit

3. Main compositions of feeding unit structure

The picture of the feeding unit is shown as in Figure 5-15. Its main compositions include workpiece loading pipe, workpiece pushing device, support frame, valve block, terminal block assembly, PLC, emergency stop button and start/stop button, wiring trough, bottom plate, etc.

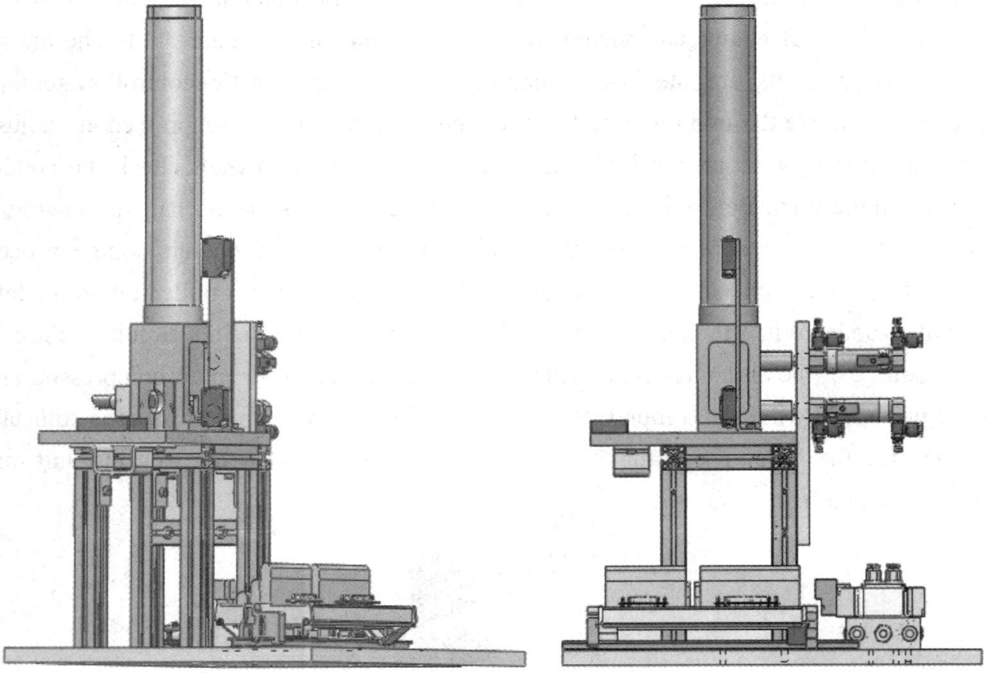

Figure 5-15 The main structure of the feeding unit

Among them, the workpiece loading pipe and the workpiece pushing device are used to store the raw material, push the lowermost workpiece from the silo to the material table when needed. The workpiece pushing device is mainly composed of a tube-shaped silo, a pusher cylinder, an ejector cylinder, a magnetic induction proximity switch and a diffuse photoelectric switch sensor.

The working principle of the workpiece pushing device is as follows. The workpiece is stacked vertically in the silo, and the pusher cylinder is at the bottom of the silo and its piston rod can pass through the bottom of the silo. When the piston rod is in the retracted position, it is at the same horizontal position as the lowermost workpiece, and the clamping cylinder is at the same horizontal position as the next lower workpiece. When the workpiece needs to be pushed out onto the material table, the piston rod of the clamping cylinder is first pushed out to press the next lower workpiece; then the piston rod of the pusher cylinder is pushed out to push the lowermost workpiece onto the material table. After the pusher cylinder returns and is drawn out from the bottom of the silo, the clamping cylinder is returned and loosen the next lower workpiece. In this way, under the effect of gravity, the workpiece in the silo will automatically move a workpiece downward to prepare for the next ejection of the workpiece.

Install one diffuse photoelectric proximity switch on the base and one on the 4th floor of the tube-shaped silo separately. Their functions are to detect whether there is storage material in the silo or whether the storage material is sufficient. If there are no workpieces in this part, the two diffuse photoelectric proximity switches at the bottom and on the 4th layer are in the normal state. If there are only 3 workpieces on the bottom, the photoelectric proximity switch at the bottom will is in action and the switch on the 4th layer is in normal state. The normal state of the photoelectric proximity switch indicates that the workpiece is almost used up. In this way, whether there is storage material in the silo or whether the storage material is sufficient can be reflected by the signal status of the two photoelectric proximity switches.

The pusher cylinder pushes the workpiece onto the material table. There is a small hole on the material table, and a cylindrical diffuse photoelectric proximity switch is arranged under the material table, which emits light upwards when the workpiece is in working, through the small hole the light helps to detect whether there is a workpiece and to provide the system with related unit signal. In the control program of the conveying unit, the signal state can be used to determine whether the manipulator device needs to be driven to grab the workpiece. The schematic diagram of the feeding unit is shown in Figure 5-16.

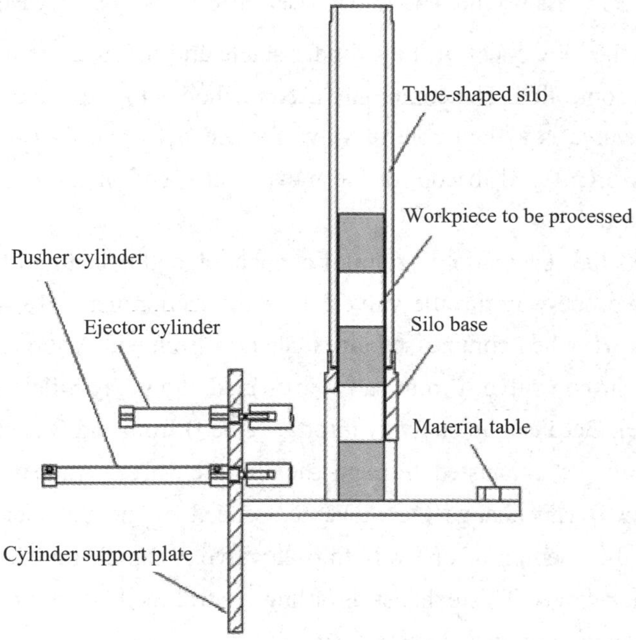

Figure 5-16 Schematic diagram of feeding unit

4. Pneumatic components of the feeding unit

1) Standard double-acting linear cylinder

Standard cylinders refer to cylinders whose functions and specifications are universally used, whose structure is easy to manufacture, and which are usually supplied as general-purpose products to the market by manufacturers.

Double-acting cylinder means that the reciprocating movement of the piston is driven by compressed air. Figure 5-17 is a half-sectional view of a standard double-acting linear cylinder. In this figure, the two end covers of the cylinder are equipped with intake port and exhaust port.

When air is taken in from the end cover port on the rodless side, the piston is pushed to move forward; conversely, when air is taken in from the end cover port on the rod side, the piston is pushed to move backward.

The double-acting cylinder has the advantages of simple structure, stable output force and the selectable stroke. However, because compressed air is alternately acted on the piston to achieve telescopic movement, the effective area of compressed air is small when retracting, so the result is that the retracting force is less than the extending force.

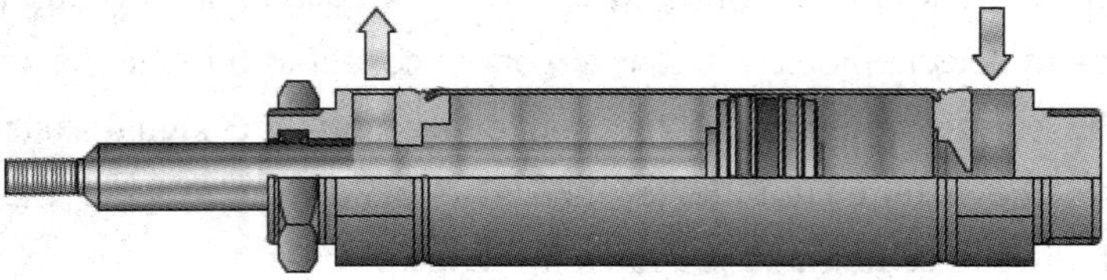

Figure 5-17 Half-sectional view of standard double-acting linear cylinder

In order to make the movement of the cylinder stable and reliable, the movement speed of the cylinder should be controlled. The commonly used method is to use a one-way throttle valve. The one-way throttle valve is a flow control valve formed by a one-way valve and a throttle valve in parallel. It is often used to control the movement speed of the cylinder, so it is also called a speed control valve.

Figure 5-18 shows the schematic diagram of connection and adjusting of the double-acting linear cylinder with two one-way throttle valves. This connection method is called the exhaust throttling method. That is, when compressed air is taken in from end A and exhausted from end B. When the one-way throttle valve A from end A is opened, it quickly inflate air into the rodless chamber of the cylinder. Because the one-way throttle valve B from end B is closed, the gas with the rod cavity can only be exhausted through the throttle valve. Adjusting the opening of one-way throttle valve B can change the movement speed of the cylinder when it extends. Conversely, adjusting the opening of one-way throttle valve A can change the movement speed of the cylinder when it retracts. This exhaust throttling control method enables the piston to run stably and is the most commonly used method.

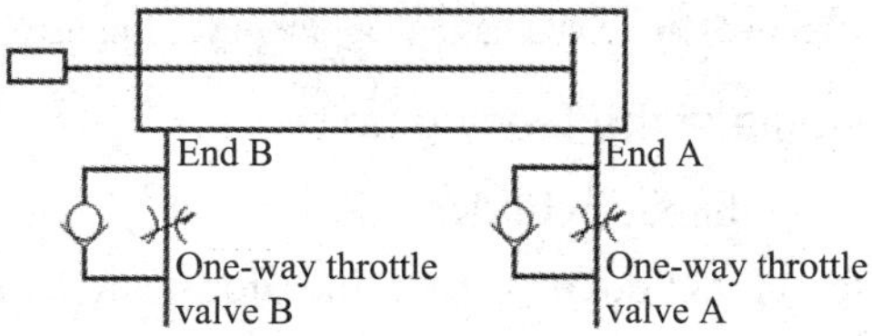

Figure 5-18 Schematic diagram of connection and adjustment of the double-acting linear cylinder with two one-way throttle valves

There is a quick connector with a trachea on the one-way throttle valve, and the pipe can be connected by just inserting a suitable outer diameter into the quick connector, which is very convenient to use. Figure 5-19 shows the appearance of a restrictive-out cylinder (with a quick coupling throttle valve).

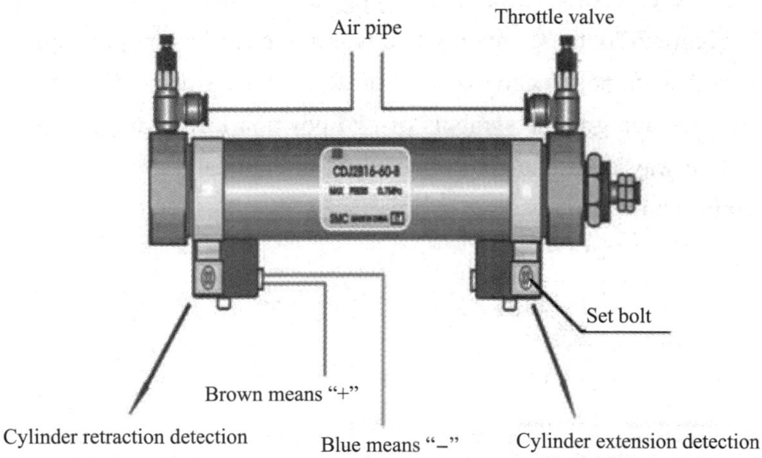

Figure 5-19 Appearance of the restrictive-out cylinder (with a quick coupling throttle valve)

2) The single electric solenoid directional control valve group and solenoid valve group

In the ejecting or pushing (pusher) cylinder, the movement of the piston is achieved by air intake from one end of the cylinder, and exhaust from the other end, and vice versa, air intake from the other end and air exhaust from the other end. The change of the gas flow direction is controlled by the control valve that can change the gas flow direction or on-off, that is, the directional control valve. In automatic control, the directional control valve often uses electromagnetic control to achieve directional control, therefore, the directional control valve is also consequently called as electromagnetic reversing valve.

The electromagnetic reversing valve uses its electromagnetic coil to energize, the static iron core produces electromagnetic attraction worked on the moving iron core to switch the valve core, achieving the purpose of changing the direction of the airflow. Figure 5-20 shows the working principle of a single electric control two-position three-way solenoid directional control valve.

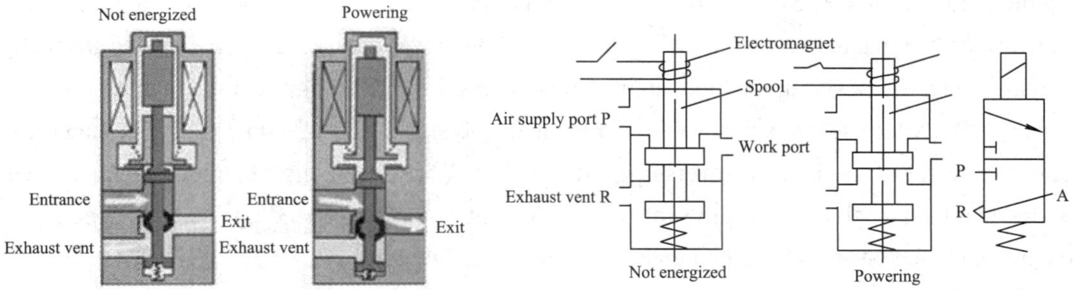

Figure 5-20 Working principle diagram of single electric control two-position three-way solenoid directional control valve

The "position" refers to the different working positions of the valve core relative to the valve body in order to change the gas direction. The number of "way" refers to the number of ports that connects the reversing valve to the system. How many ports determines how many ways they are. In Figure 5-20, there are only two working positions (not energized and powering) with air supply port P, work port A and exhaust port R, so it is a two-position three-way valve.

Figure 5-21 shows the graphic symbols of two-position three-way, two-position four-way, and two-position five-way of single-control solenoid directional valves. How many squares in the graphic determines how many ways they are.

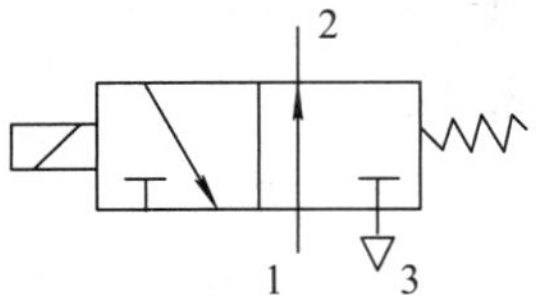

(a) Two-position three-way single-control solenoid directional valve

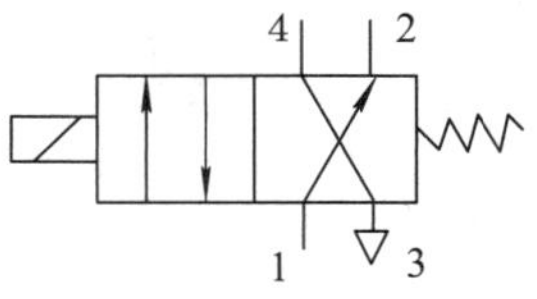

(b) Two-position four-way single-control solenoid directional valve

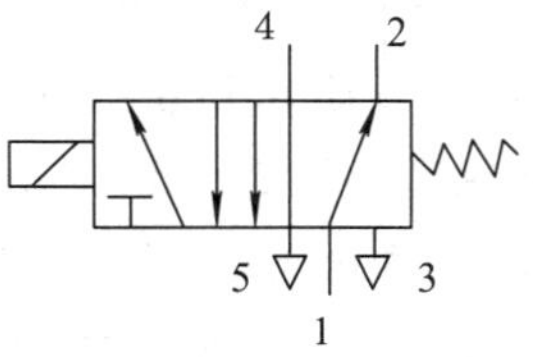

(c) Two-position five-way single-control solenoid directional valve

Figure 5-21 Graphical symbols of some single-control solenoid directional valves

The executive cylinders of all working units of YL-335B equipment are double-acting cylinders, so the solenoid valves that control their work need to have two working ports, two exhaust ports and one air supply port, the solenoid valves used are all two-position five-way solenoid valves.

The feeding unit uses two two-position five-way single-control solenoid valves. These two solenoid valves are equipped with manual reversing switch (locking button), the button has two positions: LOCK and PUSH. When the locking button is turned to the LOCK position with a small screwdriver, the manual reversing switch is recessed downward and cannot be operated manually. Only in the PUSH position, the tool can be used to press down. The signal of the manual reversing switch is "1", which is equivalent to the electromagnetic signal side of value as "1". In the normal state, the signal of the manual reversing switch is "0". When debugging the equipment, you can use the manual switch to control the valve, so as to control the corresponding gas path, specifically, to control the actuator such as the pusher cylinder for debugging.

The two solenoid valves are assembled on the manifold. Both the two ends of the exhaust ports in the manifold are connected with a muffler. The function of the muffler is to reduce the noise from the compressed air when it is discharged to the atmosphere. The integration of a group of control valves composed of multiple valves, mufflers, manifold, etc., which is called a valve group, and the function of each valve is independent of each other. The external structure of the solenoid valve group is shown in Figure 5-22.

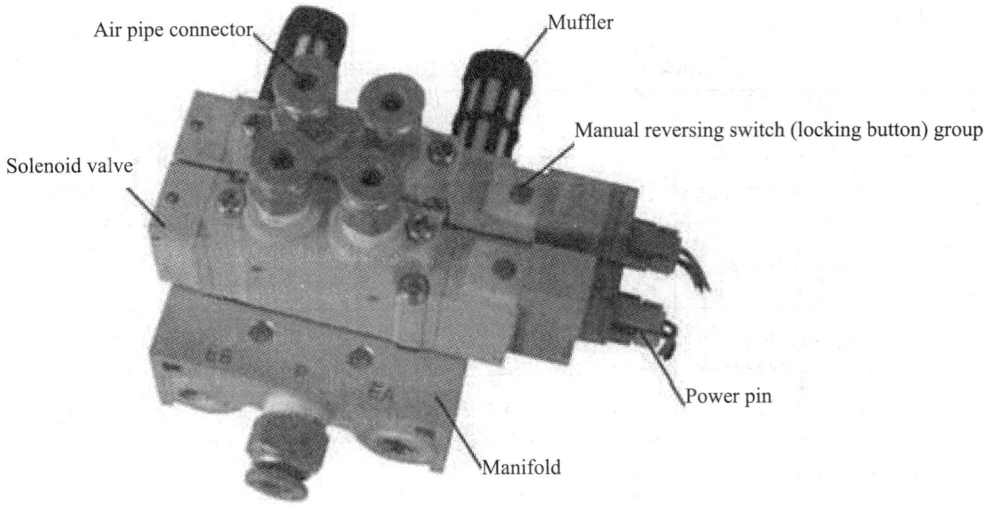

Figure 5-22 External structure of solenoid valve group

3) Linear pneumatic control circuit

The pneumatic control circuit is the actuator of the feeding unit, and the logic control function of the actuator is realized by PLC. The working principle of the pneumatic control circuit is shown in Figure 5-23. In this figure, 1A and 2A are the pusher cylinder and the ejecting cylinder respectively. 1B1 and 1B2 are magnetic induction proximity switches installed in the two extreme working positions of the pusher cylinder, and 2B1 and 2B2 are magnetic induction proximity switches installed in the two extreme working positions of the ejecting cylinder. 1Y1 and 2Y1 are the solenoid control ends of the solenoid valves that control the pusher cylinder and the ejecting cylinder, respectively. Usually, the initial positions of the two cylinders 1A and 2A are set in the retracted state.

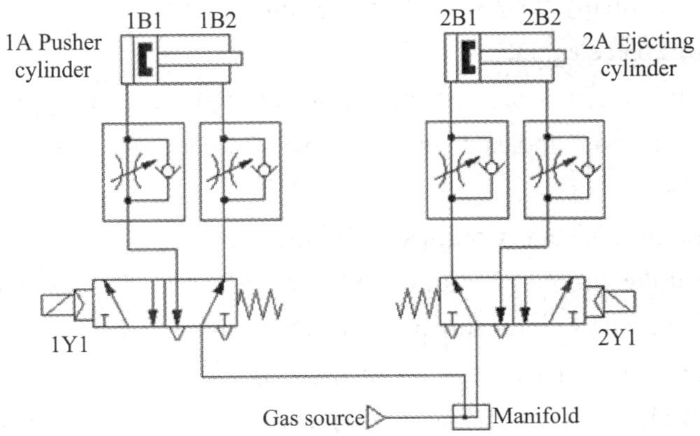

Figure 5-23 Working principle diagram of the pneumatic control circuit of the feeding unit

✍ Practice and operation

1. Task preparations

According to the requirements of this training task, prepare the required components as shown in Table 5-2.

Table 5-2 Component preparation list

No.	Name	Specification model	Amount
1	Automatic production line	YL-335B	1
2	Programming computer	Determined according to the laboratory situation	1
3	Download cable	PPI cable	1
4	Connection line	Special for YL-335B training equipment	Appropriate amount

2. Training content

The control requirements of this training task only consider the situation when the feeding unit is running as an independent device. The main signal and working status is displayed by the button/indicator module next to the PLC. The working mode selection switch SA on the button/indicator module should be placed in the "single station mode" position. The specific control requirements as follows.

(1) After the equipment is powered on and the air source is connected, if the two cylinders of the feeding unit are in the retracted position and there are enough workpieces to be processed in the silo, the "normal working" indicator light HL1 is always on, indicating the equipment get ready. Otherwise, the indicator light flashes at a frequency of 1 Hz.

(2) If the equipment is ready, press the "RUN" button to start the equipment and the "equipment operation" indicator HL2 is always on. After that, if there is no workpiece on the discharge table, the workpiece should be pushed onto the discharge table. After the workpiece on the discharge table is manually taken out, if there is no stop signal, process the second operation —pushing out the workpiece again.

(3) If the "STOP" button is pressed during operation, each feeding unit will stop working and the "equipment operation" indicator light HL2 will be dark after completing the tasks of this work cycle.

(4) If there are insufficient workpieces in the silo during operation, the feeding unit continues to work, but the "normal working" indicator light HL1 flashes at a frequency of 1 Hz, and the "equipment operation" indicator light HL2 remains on. If there is no workpiece in the silo, both the indicator light HL1 and indicator light HL2 flash at a frequency of 2 Hz, and the workstation will stop after this task cycle. Unless enough workpieces are added to the silo, the workstation can no longer be started.

The training content of this task is as follows:

(1) Plan the I/O distribution and wiring terminal distribution of the feeding unit of PLC.

(2) Perform system installation and wiring of the feeding unit.

(3) Process PLC programming according to control requirements.

(4) Debug and run the feeding unit.

3. Task procedures

1) I/O wiring of PLC

According to the I/O signal distribution of the work unit device and the requirements of the task, the feeding unit PLC selects the SMART SR40 AC/DC/RLY as main unit, which has a total of 24 points of input and 16 points of relay output. The I/O signal distribution of the feeding unit PLC is shown in Table 5-3, and its wiring schematic diagram is shown in Figure 5-24.

Table 5-3 I/O signal distribution of the feeding unit PLC

Input signal				Output signal			
No.	PLC input point	Signal name	Signal source	No.	PLC input point	Signal name	Signal source
1	I0.0	Extension detection of ejecting cylinder	Device side	1	Q0.0	Ejecting cylinder solenoid valve	Device side
2	I0.1	Retraction detection of ejecting cylinder		2	Q0.1	Pusher cylinder solenoid valve	
3	I0.2	Extension detection of pusher cylinder		3	Q0.2		
4	I0.3	Retraction detection of pusher cylinder		4	Q0.3		
5	I0.4	Material inspection on the discharge table		5	Q0.4		
6	I0.5	Material insufficient detection		6	Q0.5		
7	I0.6	Material unavailable detection		7	Q0.6		
8	I0.7	Detection of metal workpieces		8	Q0.7		
9	I1.0			9	Q1.0	Normal work indicator	Button/indicator module
10	I1.1			10	Q1.1	Operation indicator	

Continued table

Input signal				Output signal			
No.	PLC input point	Signal name	Signal source	No.	PLC input point	Signal name	Signal source
11	I1.2	Stop button	Button/indicator module				
12	I1.3	Run button					
13	I1.4						
14	I1.5	Work mode selection					

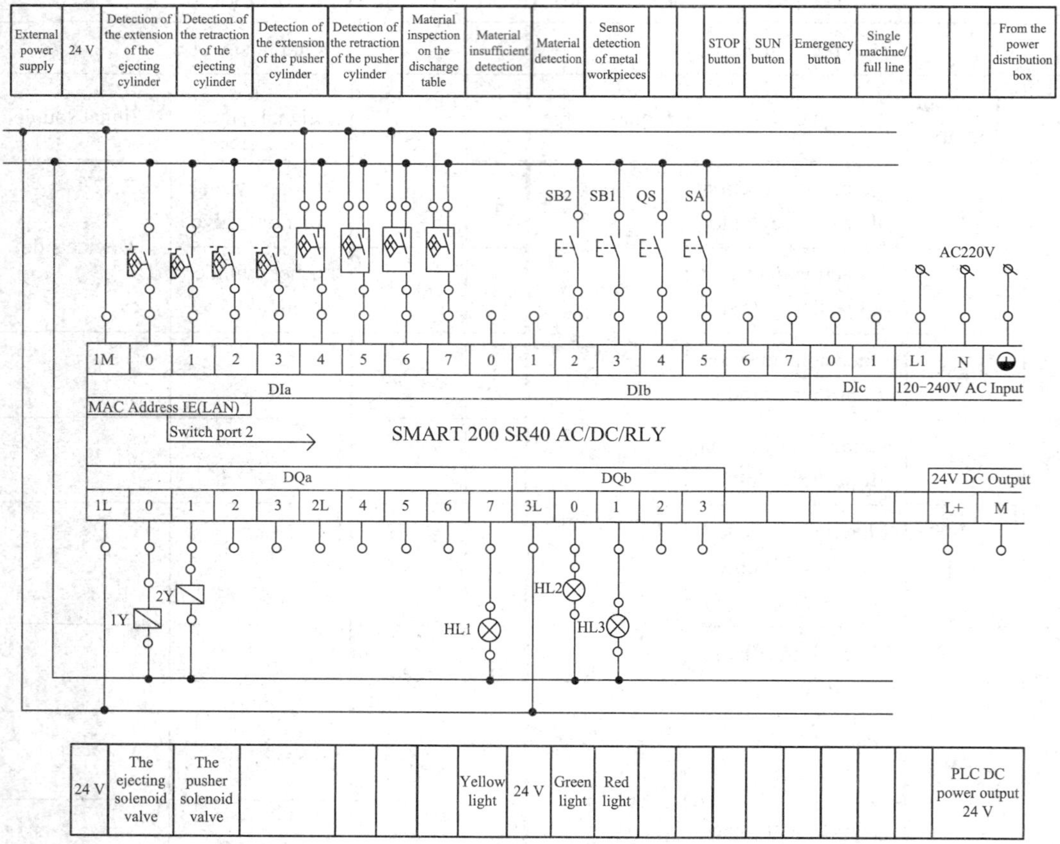

Figure 5-24　The I/O wiring schematic diagram of the feeding unit PLC

2) Programming of feeding unit control system

The structure of feeding unit control system program is written with the idea of master-slave calling. The main process of feeding unit is feeding control, which is a step-by-step sequence control process. The ladder diagram of the main program of the feeding unit control system is shown in Figure 5-25.

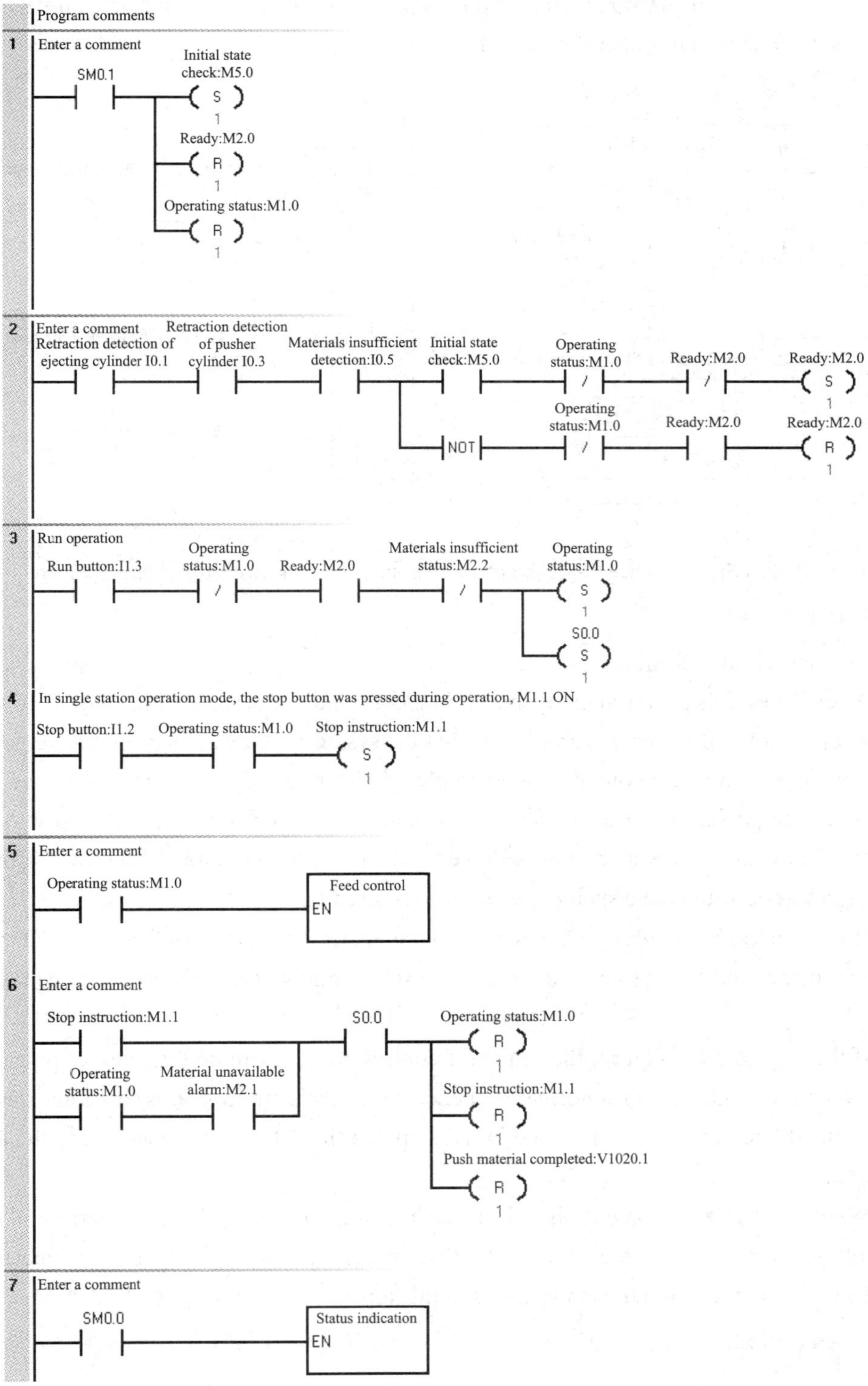

Figure 5-25 Ladder diagram of the main program of the feeding unit control system

The step sequence flow chart of the subroutine of the feed control system program is shown in Figure 5-26 below. In the figure, the initial step S0.0 is set in the main program when the system is ready and the start pulse is received.

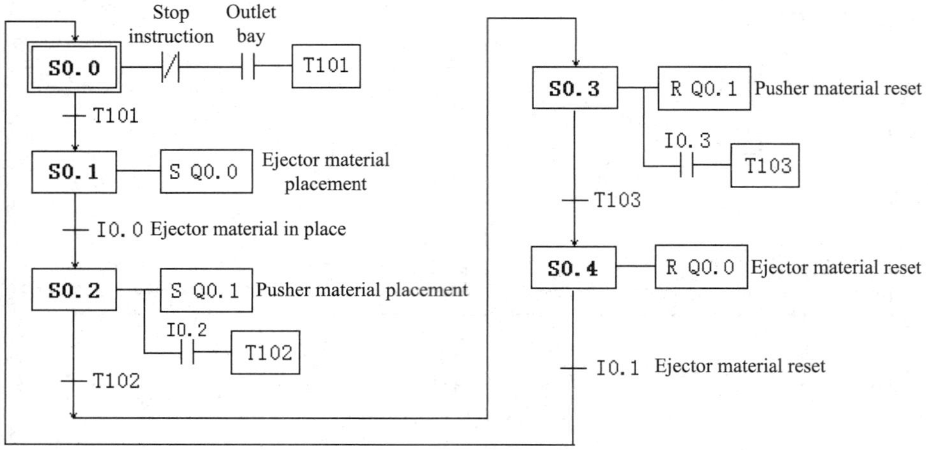

Figure 5-26　Step sequence flow chart of the subroutine of the feed control system program

4. Precautions

Precautions are as follows.

(1) After the PLC is powered on, it should first enter the initial inspection stage. The system be put into operation only after it is confirmed that the system is ready, can problems be found in time and accidents can be avoided. For example, if the two cylinders are not in the initial position when the power is turned on and the gas source is connected, this is due to wrong gas connection. Obviously, the system is not allowed to be put into operation in this case. Common PLC control systems often have such conventional requirements.

(2) When compiling a program, the main program calls the system to display the subroutine in each scan cycle, and the system calls to use the subroutine when the running status has been established.

(3) If there is no stop request, the sequence control process will continue to loop again and again. A common procedure for a normal sequence control system to stop is that after receiving a stop command, the system returns to the initial step and then stops after completing the task of this work cycle.

(4) When the last workpiece in the silo is pushed out, a material shortage alarm will occur. The pusher cylinder is reset to the position, that is, the task of this work cycle is completed, and the initial step is returned, and the entire task should stop.

5. Notes and summary

Task 2 Programming and debugging of processing unit control system

✍ Relevant knowledge

1. Processing unit structure and function

The function of the processing unit is to transfer the workpiece to be processed from the material table to just below the punching cylinder in the processing area, completing the punching of the workpiece, and then returning the processed workpiece to the material table.

The main structural components of the processing unit are processing table and sliding mechanical structure (introduced in this task), processing (punching) mechanical structure (introduced in this task), solenoid valve block, wiring port, bottom plate, etc. The mechanical structure assembly of the processing unit is shown in Figure 5-27.

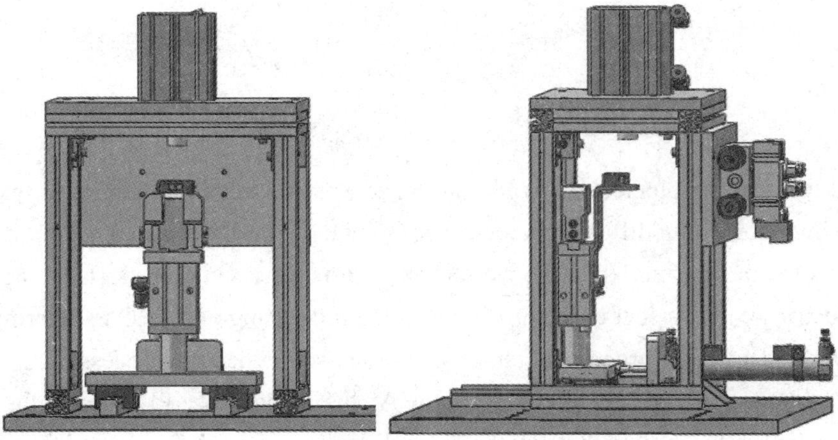

(a) Front view (b) Rear view

Figure 5-27 Assembly diagram of the mechanical structure of the processing unit

1) Processing table

The processing table is shown in Figure 5-28. The sliding processing table is used to fix the workpiece and move the workpiece directly to the position under the sliding mechanical structure for punching processing. It is mainly composed of pneumatic claw, pneumatic fingers, processing table, telescopic cylinders, linear guides and sliders, magnetic induction proximity switches and diffuse photoelectric switches.

The working principle of the processing table: the system is in normal working state. the initial state of the processing table is that the telescopic cylinder in it is extended, and the pneumatic fingers of the processing table are opened. When the sliding mechanical structure sends the material to the material table, the material detection sensor detects the workpiece. After that, the PLC control program completes the workpiece processing procedures as follows: drive the pneumatic fingers to clamp the workpiece → the processing table returns to the processing place under the punching cylinder → the piston rod of the punching cylinder extends downwards to punch the workpiece → the piston rod retracts upwards after the punching action is completed → the processing table is extended again → the pneumatic fingers are released → finish the processing. And then the program sends a processing completion signal to the system and prepare for the next time the workpiece arrives for processing.

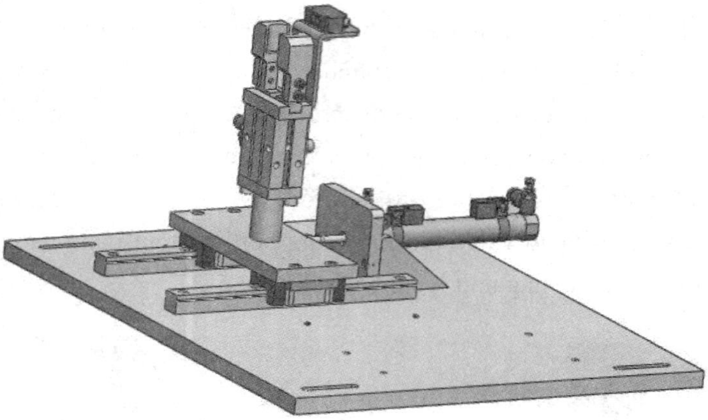

Figure 5-28　Processing table

Install a diffuse photoelectric switch on the processing table. If there is no workpiece on the processing table, the diffuse photoelectric switch is in the normal state; if there is a workpiece on the processing table, the photoelectric proximity switch acts. The output signal of the photoelectric switch is sent to the input end of the processing unit PLC to determine whether there is a workpiece on the processing table to be processed; when the processing is finished, the processing table extends to the initial position. At the same time, PLC feeds the processing completion signal to the system through the communication network for control coordination.

The diffuse photoelectric switch installed on the processing table uses the E3Z-L63 amplifier built-in photoelectric switch (small beam type). The principle, structure and debugging method of the photoelectric switch have been introduced before.

The position to which the material table extends and returns is positioned by the position adjustment of the two magnetic switches on the telescopic cylinder. The retracted position is required to be directly below the punching head; the extended position should be matched with the grabbing manipulator of the conveying unit to ensure that the grabbing manipulator of the conveying unit can smoothly put the workpiece to be processed on the material table.

2) Processing (punching) mechanical structure

The processing (punching) mechanical structure (aka sliding mechanical structure) is shown in Figure 5-29. It is used for punching the workpiece. It is mainly composed of punching cylinder, punching head, mounting plate and so on.

The working principle of the punching table is as follows: when the workpiece reaches the punching position, that is, the piston rod of the telescopic cylinder is retracted, and the punching cylinder extends to process the workpiece. After the processing is completed, the punching cylinder retracts to prepare for the next punching.

According to the requirements of the workpiece, and the punching head is installed on the surface of the punching cylinder. The mounting plate is used to install and fix the punching cylinder.·

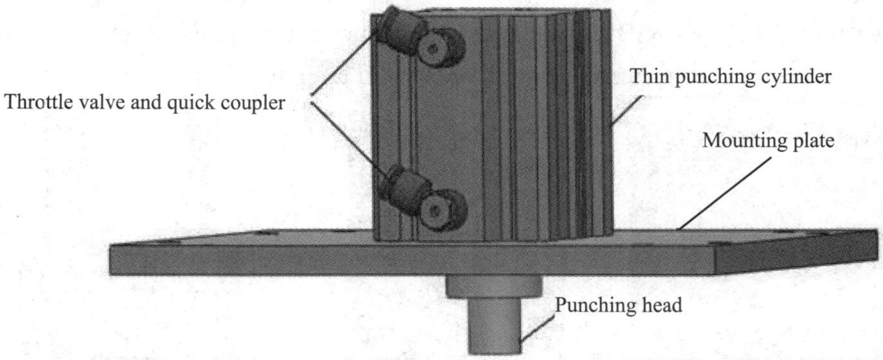

图 5-29 Punching mechanical structure

2. Pneumatic actuators of the processing unit

The pneumatic actuators used in the processing unit include standard linear cylinder, thin cylinder and pneumatic finger. Linear cylinders have been introduced in Task 1 in this module. Only thin cylinder and pneumatic finger are introduced below.

1) Thin cylinder

Thin cylinder belong to the category of space-saving cylinders, that is, cylinder whose axial or radial dimensions are much smaller than standard cylinder. It has the advantages of compact structure, light weight, and small space occupation. Figures 5-30 are examples of thin cylinder.

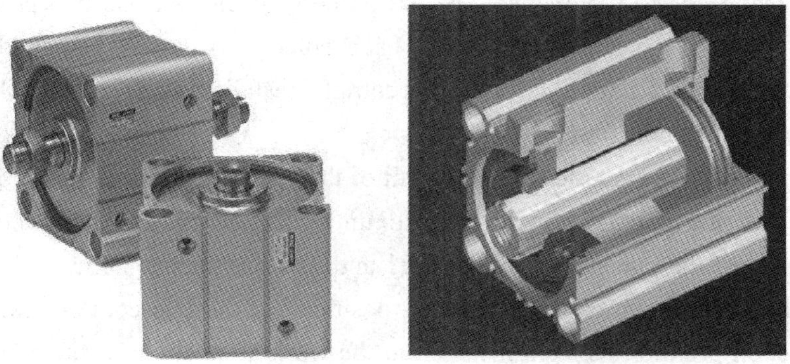

(a) Pictures of thin cylinder (b) Section view of thin cylinder

Figure 5-30 Pictures and section view of thin cylinder

The characteristics of the thin cylinder are as follows: the cylinder barrel and the end cover on rodless side are die-casted into one body, the rod cover is fixed with an elastic retaining ring, and the cylinder body is square. This kind of cylinder is usually used as fixtures (fixing workpieces during transportation). In the YL-335B processing unit, the thin cylinder is used for punching because of its short stroke.

2) Pneumatic fingers (air grippers)

The air grippers are used to grab and clamp the workpiece. The working modes of pneumatic fingers usually have sliding guideway type, pivot opening and closing type, and rotary drive type. The YL-335B equipment's pneumatic finger is sliding guideway type. The section view and picture of pneumatic fingers are shown in Figure 5-31.

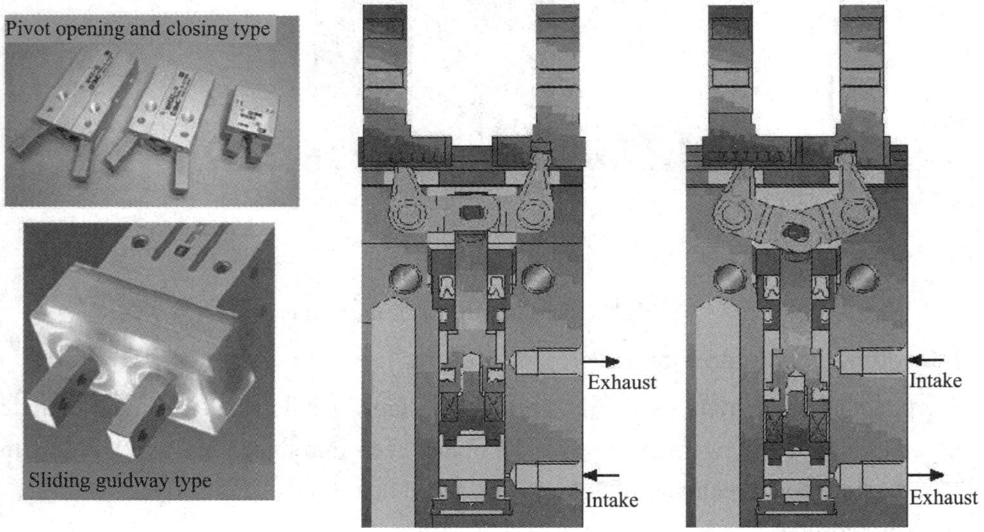

(a) Picture of Air grippers (b) Section view in loosened state (c) Section view in clamped state

Figure 5-31 Picture and section view of pneumatic fingers

3. Pneumatic control circuit

The pneumatic control components of the processing unit all use two-position five-way single-control solenoid reversing valve, and each solenoid valve is equipped with manual reversing switch (locking button). They are centrally installed as the valve block and fixed behind the punching support frame.

The schema of the pneumatic control circuit of the processing unit is shown in Figure 5-32. 1B1 and 1B2 are the magnetic induction proximity switches installed in the two extreme working positions of the punching cylinder. 2B1 and 2B2 are the magnetic induction proximity switches installed in the two extreme working positions of the telescopic cylinder. 3B1 is the magnetic induction proximity switch installed in the working position of the material clamping cylinder. 1Y1, 2Y1, and 3Y1 are the electromagnetic control ends of the solenoid valves that control the punching cylinder, the telescopic cylinder, and the material clamping cylinder, respectively.

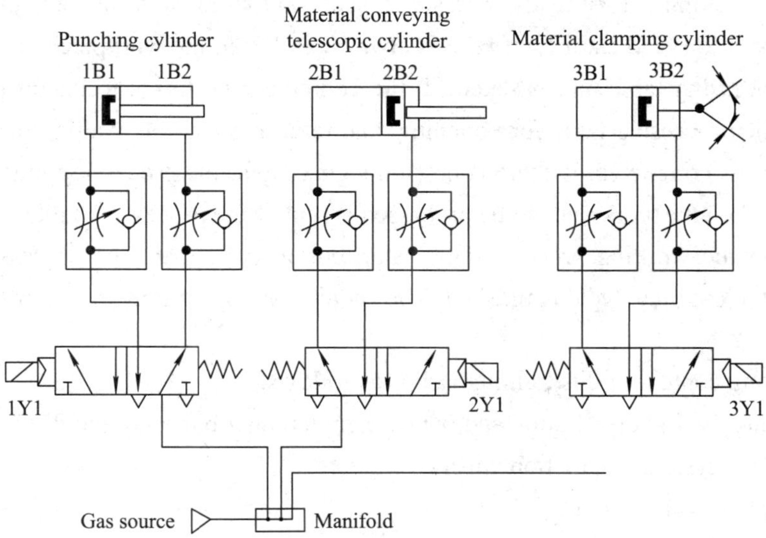

Figure 5-32 Schema of the pneumatic control circuit of the processing unit

✍ Practice and operation

1. Task preparations

According to the requirements of this training task, prepare the required components as shown in Table 5-4.

Table 5-4 Component list

No.	Name	Specification model	Amount
1	Automatic production line	YL-335B	1
2	Programming computer	Determined according to the training situation	1
3	Download cable	PPI cable	1
4	Connection line	YL-335B special for training equipment	Appropriate amount

2. Training content

The control requirements of this training task only consider the situation when the processing unit is used as an independent device. The working mode selection switch of the button/indicator unit of this module should be placed in the “single station mode” position. The specific control requirements are as follows.

(1) Initial state. After the equipment is powered on and the air source is connected, the telescopic cylinder of the processing table is in the extended position; the pneumatic fingers of the processing table is released; the punching cylinder is in the retracted position; and the emergency stop button is not pressed. If the device is in the above initial state, the “normal working” indicator light HL1 is always on, indicating that the device is ready. Otherwise, the indicator light flashes at a frequency of 1 Hz.

(2) If the equipment is ready, press the start button to start the equipment, and the "equipment operation" indicator light HL2 is always on. When the workpiece to be processed is sent to the processing table and is detected, the equipment performs to clamp the workpiece, sending it to the processing table for punching, and returning it to the waiting position after the punching action is completed. If there is no stop signal input, the processing unit starts the next cycle of work when the workpiece to be processed is sent to the processing table.

(3) During the working process, if the Stop button is pressed, the processing unit stops working after the ongoing cycle is finished. The "equipment operation" indicator light HL2 goes out.

The working content of this training task is as follows.

(1) Planning the I/O distribution and wiring terminal distribution of the PLC.

(2) Carry out system installation wiring.

(3) Compile PLC program.

(4) Debug and run.

3. Task procedures

1) I/O wiring of the feeding unit of PLC

The feeding unit of PLC uses SMART200 SR40 AC/DC/RLY as main unit with a total of 24 inputs and 16 relay outputs. The I/O signal distribution table of the processing unit PLC is shown in Table 5-5, and the wiring schematic diagram is shown in Figure 5-33.

Table 5-5 I/O signal table of the processing unit PLC

Signal type	Signal parameters				Signal type	Signal parameters			
	No.	PLC input point	Signal name	Signal source		No.	PLC output point	Signal name	Signal source
Input signal	1	I0.0	Processing table material inspection	Device side	Output signal	1	Q0.0	Clamping solenoid valve	Device side
	2	I0.1	Workpiece clamping detection			2	Q0.1		
	3	I0.2	Detection of the extension of the processing table			3	Q0.2	Telescopic solenoid valve	
	4	I0.3	Detection of the retraction of the processing table			4	Q0.3	Pneumatic actuator solenoid valve	
	5	I0.4	Upper limit of pneumatic actuator detection			5	Q0.4		
	6	I0.5	Lower limit of pneumatic actuator detection			6	Q0.5		
	7	I0.6				7	Q0.6		

Continued table

Signal type	Signal parameters				Signal type	Signal parameters			
	No.	PLC input point	Signal name	Signal source		No.	PLC output point	Signal name	Signal source
Input signal	8	I0.7			Output signal	8	Q0.7		
	9	I1.0				9	Q1.0	Normal work instructions	Button/ indicator module
	10	I1.1				10	Q1.1	Operation instructions	
	11	I1.2	Stop button	Button/ indicator module					
	12	I1.3	Run button						
	13	I1.4	Emergency button						
	14	I1.5	Single station/full line						

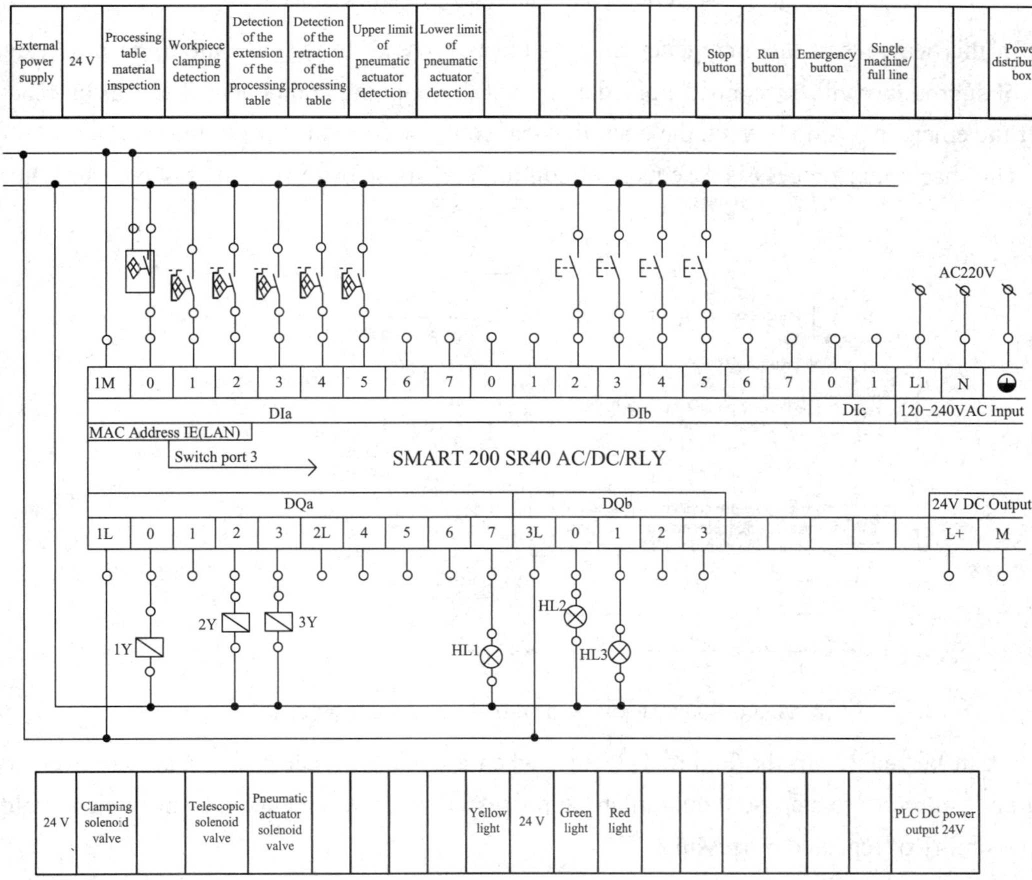

Figure 5-33 I/O wiring schematic diagram of the processing unit PLC

2) Programming and debugging of PLC control program

The main program flow of the processing unit is similar to that of the feeding unit. After the PLC is powered on, it should first enter the initial state check stage. After confirming that the system is ready, it is allowed to receive the start signal and put it into operation. However, an emergency stop function is added to the task of the machining unit. For this reason, the condition for calling the processing control subroutine should be that both "unit is running" and "emergency stop button not pressed" are both established at the same time. Invoking of the processing control subroutine is shown in Figure 5-34.

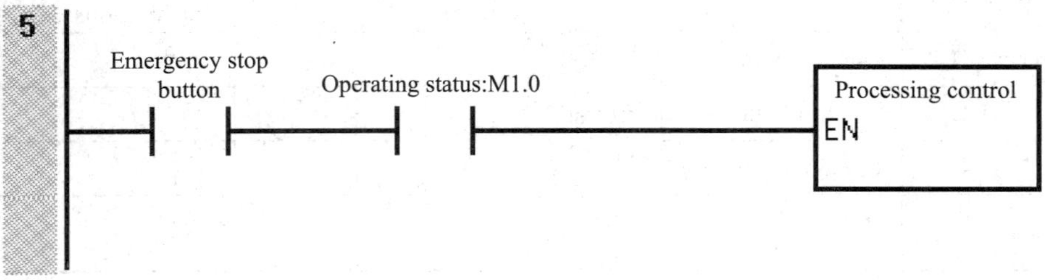

Figure 5-34 Calling of the processing control subroutine

In this way, when the emergency stop button is pressed during operation, the processing control subroutine will be stopped immediately, but the ongoing component S is still on-ready. After the emergency stop is reset, the operation can continue from the breakpoint.

The machining process is a sequential control, and its step flow chart is shown in Figure 5-35.

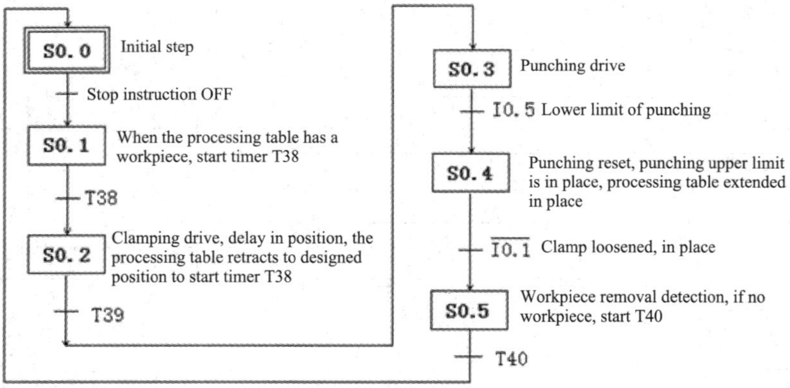

Figure 5-35 Step flow chart of the machining process

It can be seen from the Figure 5-35 that when a cycle of machining process is over, the program can return to step S0.0 only after the processed workpiece is taken away, which avoids the possibility of repeated processing.

4. Precautions

There are some precautions during programming:

(1) It is necessary to ensure that the PLC should enter the initial statue checking stage first after power-on. Only after we confirm that the system is ready, can it be allowed to receive the start signal and put it into operation.

(2) The program should be connected to the emergency stop button, ensure that when the emergency stop button is pressed, the system will immediately stop invoking the processing control subroutine. After the emergency stop is reset, continue to run the processing program from the breakpoint.

(3) The program should ensure that the processed workpiece is taken away before entering the next processing cycle.

5. Notes and summary

____

____

____

____

____

____

____

Task 3 Programming and debugging of assembly unit control system

✍ Relevant knowledge

The function of the assembly unit is to embed the small black or white metal cylindrical parts from the unit silo to the workpiece assembly hopper.

The structural composition of the assembly unit includes tubular silo, blanking mechanical structure, material turntable, assembly manipulator, assembly table hopper, warning light, positioning mechanical structure of the workpiece to be assembled, pneumatic system and its valve group, signal acquisition and automatic control system, the terminal block assembly used for electrical connection, the signal light indicating the status of the entire production line, the aluminum profile bracket and the bottom plate for the installation of other mechanical structure, and the sensor mounting bracket and other accessories. The mechanical structure drawing of assembly unit is shown in Figure 5-36.

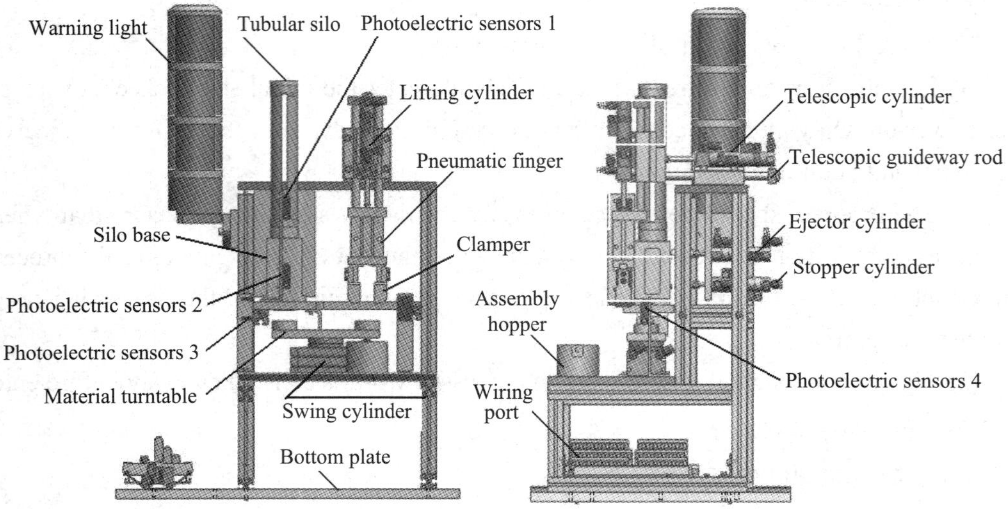

Figure 5-36 Mechanical structure drawing of the assembly unit

1. Tubular silo

The tubular silo is used to store the small black or white metal cylindrical parts for assembly. It is composed of a hollow plastic round tube and a hollow base. A reinforced metal ring is placed on the top of the plastic round tube to prevent damage. The workpiece is placed vertically in the hollow plastic round tube of the silo. Due to a certain gap between the two, it can fall freely under the action of gravity.

In order to alarm when the silo is insufficient, two diffuse reflection photoelectric sensors (type E3Z-L) are installed at the bottom of the hollow plastic round tube and the hollow base respectively, and a slot is milled longitudinally on the hollow plastic round tube of the silo, in order that the infrared light spot can reliably irradiate on the material to be detected by the photoelectric sensor, and the photoelectric sensor should be adjusted to keep the sensitivity when it can detect black materials.

2. Blanking mechanical structure

Figure 5-37 shows a cross-sectional view of the blanking mechanical structure. In this figure, two linear cylinders are installed on the back of the silo base. The upper cylinder is called the ejector cylinder, and the lower cylinder is called the stopper cylinder.

After the air source is connected, the initial position of the ejector cylinder is in the retracted state, and the initial position of the stopper cylinder is in the extended state. In this way, when the workpiece is put down from the silo, the workpiece will be blocked by the stopper at the end of the piston rod of the stopper cylinder and cannot fall.

When the blanking operation is required, the ejector cylinder is firstly extended to clamp the next lower workpiece. Then the stopper cylinder is retracted, and the workpiece falls into the material tray of the material turntable. After that, the stopper cylinder is reset and extended, and the ejector cylinder is retracted. The next lower workpiece falls onto the end stop of the stopper cylinder to prepare for blanking again.

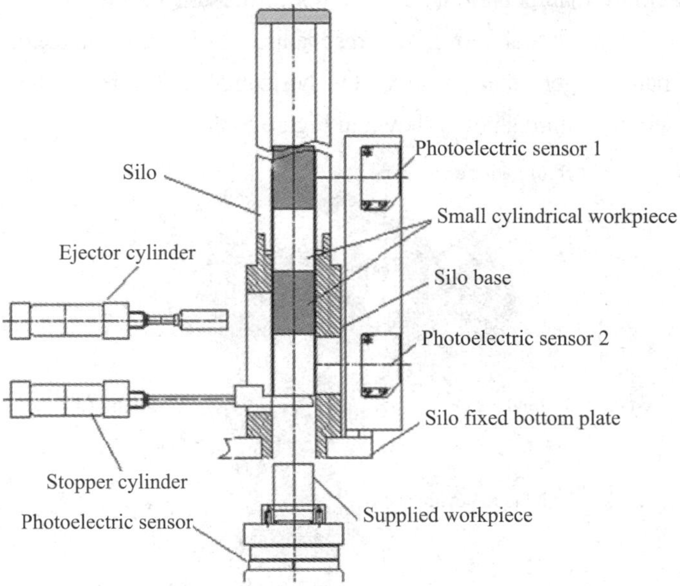

Figure 5-37 Sectional view of blanking mechanical structure

3. Material turntable

The material turntable is composed of a swing cylinder and two material trays. The swing cylinder can drive the material tray to rotate 180 degrees, so as to realize the function of moving the workpiece falling from the feeding mechanical structure to the material tray and then to the place which is directly below the assembly manipulator. The structure of the material turntable is shown in Figure 5-38. The photoelectric sensor 1 and photoelectric sensor 2 in this figure are respectively used to detect whether there are parts on the left and the right material trays. Both photoelectric sensors use CX-441 type.

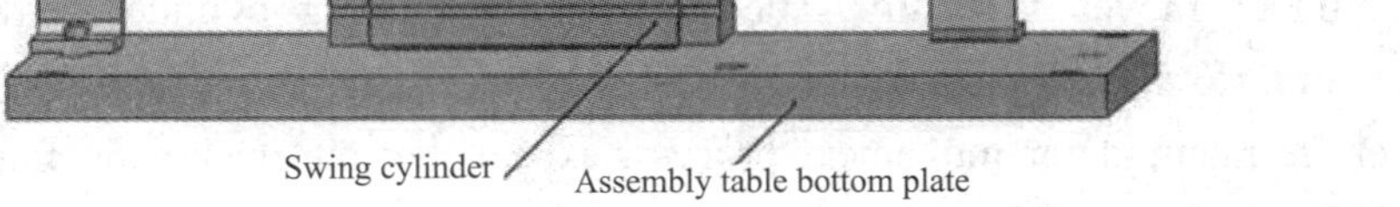

Figure 5-38 The structure of the material turntable

4. Assembly manipulator

The assembly manipulator is the core of the entire assembly unit. When there are small cylindrical parts on the material tray (the tray is above the material turntable, the turntable is directly below the assembly manipulator), and the photoelectric sensor on the side of the assembly table detects a workpiece, the assembly manipulator starts operation process from the initial state. The assembly manipulator is a three-dimensional movement mechanical structure, which consists of two guide cylinders and pneumatic fingers that move in the horizontal and vertical directions. The overall appearance of the assembly manipulator is shown in Figure 5-39.

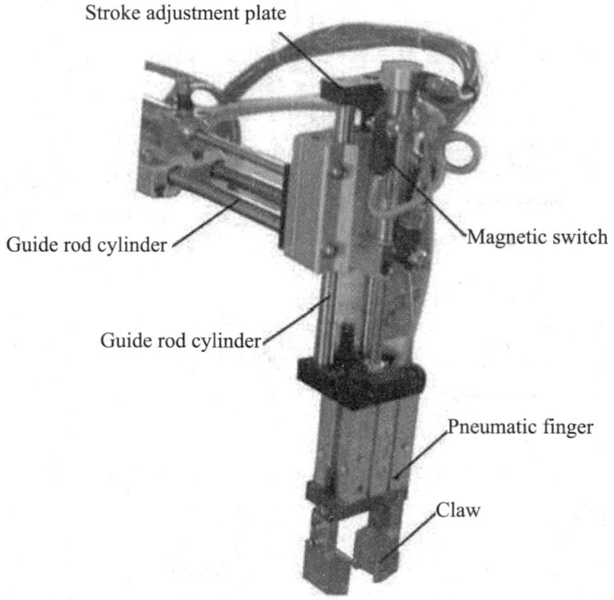

Figure 5-39 Overall drawing of the assembly manipulator

The operation process of the assembly manipulator is as follows. The PLC drives the solenoid reversing valve connected with the vertically moving cylinder to move. The pneumatic finger is driven by the vertically moving cylinder with a guide rod to move down. When in place, the pneumatic finger drives the claw to clamp the material, and the clamping signal passes through the magnetic switch and be transfered to the PLC. Under the control of the PLC, the vertical moving cylinder is reset. The clamped material is lifted up with the pneumatic finger and left when the material tray of the turning material table is lifted to the highest position. The horizontal moving cylinder is corresponding to it. Driven by the reversing valve, the piston rod extends and moves to the front end of the cylinder. The vertically moving cylinder is driven down again to move to the lowest position. The pneumatic fingers are released. After a short delay, the cylinder moves vertically. The horizontally moving cylinder retracts, and the manipulator returns to the initial state.

In the entire manipulator movement process, except that the pneumatic finger release in place is unable to be detected by sensor, the detection of the in-position signal of other actions adopts the magnetic switch matched with the cylinder, the working process of the magnetic switch is as follows: the collected signal is input to the PLC to get the PLC output signal which drives the solenoid valve to change direction, so the manipulator (composed of cylinder and pneumatic fingers) run automatically according to the program.

5. Assembly table hopper

The workpiece to be assembled transported by the conveying unit is directly placed in the positioning hole of the hopper, and the positioning is realized by the small clearance fit between the positioning hole and the workpiece, thereby accurate assembly action and accuracy positioning is completed. The assembly table hopper is shown in Figure 5-40.

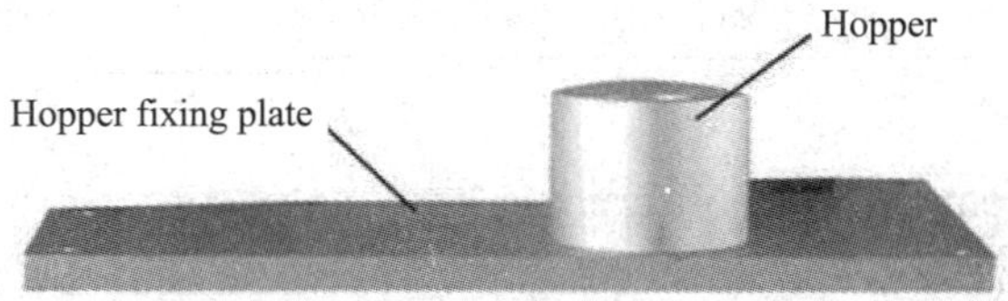

Figure 5-40 Assembly table hopper

In order to determine whether the workpiece to be assembled is placed in the hopper of the assembly table, an optical fiber sensor is used for detection. An M6 screw hole is opened on the side of the hopper, and the optical fiber probe of the photoelectric sensor is fixed in the M6 screw hole.

6. Warning Light

Red, orange, and green three-color warning lights are installed on this working unit, which are used as warnings for the entire system. The warning light has five lead wires, among which the yellow-green cross wire is the "ground wire"; the red wire is the red light control wire; the yellow wire is the yellow light control wire, the green wire is the green light control wire; the black wire is the public control wire of signal light. The warning lights and wiring are shown in Figure 5-41.

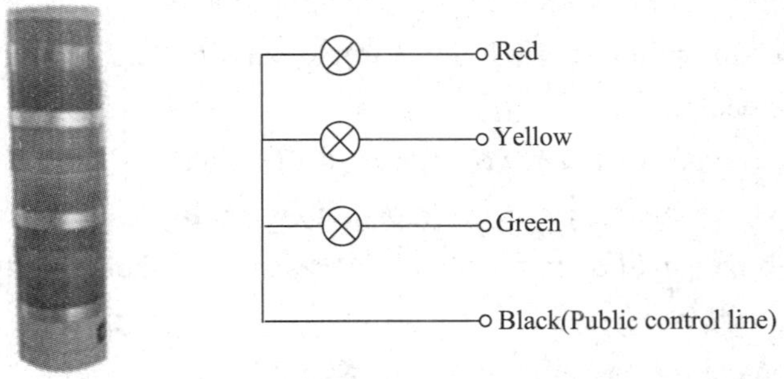

(a) Warning light shape (b) Warning light wiring principle

Figure 5-41 Warning light and its wiring

✍ Practice and operation

1. Task preparations

According to the requirements of this training content, prepare the required components as shown in Table 5-6 below.

Table 5-6 Component Preparation List

No.	Name	Specification model	Amount
1	Automatic production line	YL-335B	1
2	Programming computer	Determined according to the laboratory situation	1
3	Download cable	PPI cable	1
4	Connection lines	Special YL-335B lines for training equipment	Appropriate amount

2. Training content

The control requirements of this training task only consider the situation when the processing unit is used as an independent machine. The working mode selection switch on the button/indicator unit of this module should be placed in the "single station mode" position. The specific control requirements are:

(1) The initial position of each cylinder of the assembly unit is as follows. The stopper cylinder is in the extended state, and the ejector cylinder is in the retracted state. There should be enough small cylinder workpieces on the silo, and the lifting cylinder of the assembly manipulator is in the raised state, the telescoping cylinder is in the retracted state, and the pneumatic gripper is in the released state. After the equipment is powered on and the air source is connected, if each cylinder meets the initial position requirements, there are enough small cylindrical parts on the silo, and there is no workpiece to be assembled on the workpiece assembly table. The "normal working" indicator HL1 is always on, indicating that the device is ready. Otherwise, the indicator light flashes at a frequency of 1 Hz.

(2) If the equipment is ready, press the start button to start the assembly unit, and the "equipment operation" indicator HL2 is always on. If there are no small cylinder workpieces (parts) in the left tray of the turntable, the blanking operation will be executed. If there are cylinder workpieces in the left tray but there are no parts in the right tray, the turntable rotation operation will be executed.

(3) If there are small cylinder workpieces in the right tray on the turntable and there are workpieces to be assembled on the assembly table, perform the operation of the assembly manipulator to grab the small cylindrical workpieces and put them into the assembly table to be assembled.

(4) After complete the assembly task, the assembly manipulator should return to the initial position and wait for the next assembly.

(5) If the stop button is pressed during operation, the blanking mechanical structure should stop feeding immediately. If the assembly conditions are met, the assembly unit will stop working until this assembly is completed.

(6) When the "workpieces shortage" alarm occurs during operation, the indicator light HL3 flashes at a frequency of 1 Hz, and the HL1 and HL2 lights are always on. When the "workpieces missing" alarm occurs during operation, the indicator light HL3 is on for 1 second and then off flashes for 0.5 seconds, HL2 is off, and HL1 is always on.

The requirements of the working content of this training task is as follows.

(1) Process the I/O distribution and wiring terminal distribution of the PLC.

(2) Carry out system installation wiring and gas circuit connection.

(3) Compile PLC program.

(4) Debug and run.

3. Task procedures

1) PLC I/O distribution and system installation wiring

The assembly unit has many I/O points, and the SMART200 SR40 AC/DC/RLY is selected as main unit, with a total of 24 points of input and 16 points of relay output. PLC I/O allocation is shown in Table 5-7. Figure 5-42 is the schematic diagram of the assembly unit PLC wiring.

Table 5-7 I/O allocation of the assembly unit PLC

Input signal				Output signal			
No.	PLC input point	Signal name	Signal source	No.	PLC input point	Signal name	Signal source
1	I0.0	Parts insufficient detection	Device side	1	Q0.0	Stopper solenoid valve	Device side
2	I0.1	Parts unavailable detection		2	Q0.1	Ejector solenoid valve	
3	I0.2	Left tray parts detection		3	Q0.2	Rotary solenoid valve	
4	I0.3	Right tray parts detection		4	Q0.3	Claw clamping solenoid valve	
5	I0.4	Assembly table workpiece detection		5	Q0.4	Claw down solenoid valve	
6	I0.5	Ejector in place detection		6	Q0.5	Arm out solenoid valve	
7	I0.6	Ejector reset detection		7	Q0.6	Red warning light	
8	I0.7	Stopper cylinder detection		8	Q0.7	Orange warning light	
9	I1.0	Blanking status detection		9	Q1.0	Green warning light	
10	I1.1	Rotary cylinder left limit detection		10	Q1.1		
11	I1.2	Rotary cylinder right limit detection		11	Q1.2		
12	I1.3	Claw clamping detection		12	Q1.3		
13	I1.4	Claw dropping to position detection		13	Q1.4		
14	I1.5	Claw rising to position detection		14	Q1.5	HL1	Button/ indicator module
15	I1.6	Claw retracted to position detection		15	Q1.6	HL2	
16	I1.7	Claw extended to position detection		16	Q1.7	HL3	
17	I2.0						
18	I2.1						

Continued table

Input signal				Output signal			
No.	PLC input point	Signal name	Signal source	No.	PLC input point	Signal name	Signal source
19	I2.2						
20	I2.3						
21	I2.4	Stop button	Button/ indicator module				
22	I2.5	Run button					
23	I2.6	Emergency button					
24	I2.7	Stand-alone/online					

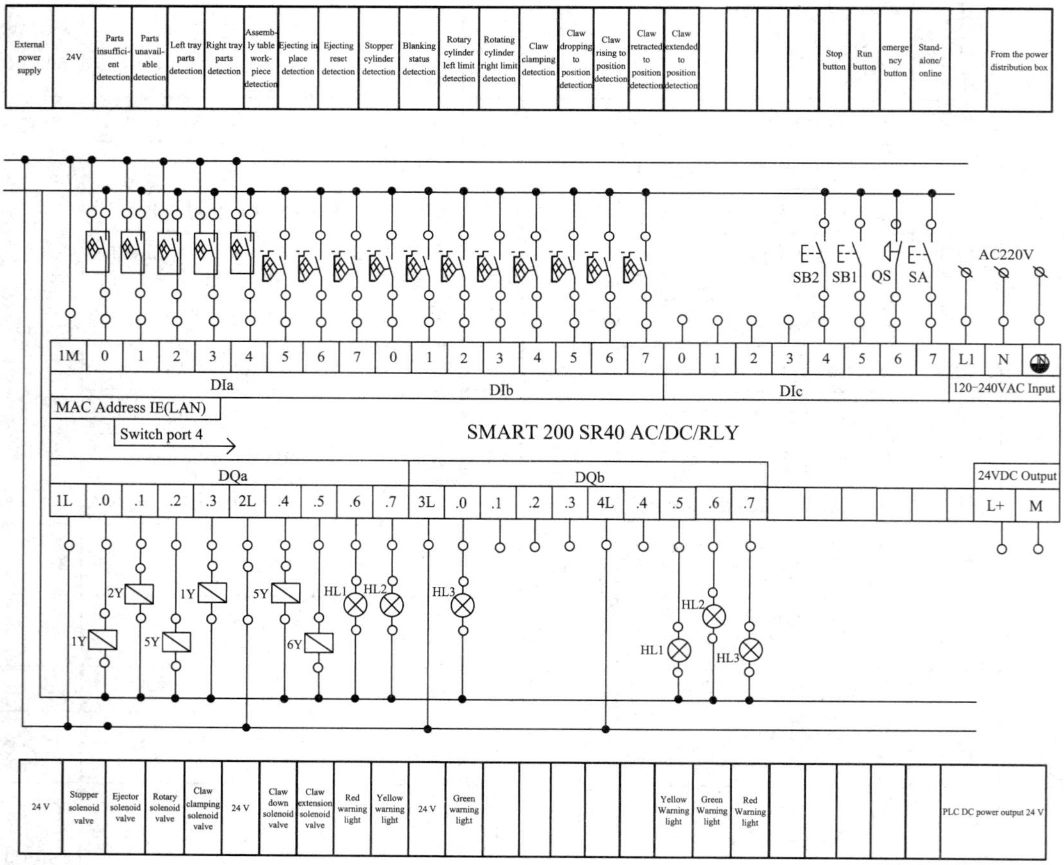

Figure 5-42　Schematic diagram of the assembly unit PLC wiring

2) PLC control program programming and debugging

After entering the running state, the working process of the assembly unit includes two mutually independent sub-processes: one is the blanking process and the other is the assembly process.

(1) The blanking process is to make the small cylindrical workpieces in the silo fall onto the tray on the left side of the swing table through the operation of the blanking mechanism; then the swing table rotates to transfer the tray with the workpieces to the right for the assembly manipulator to grab cylindrical workpieces.

(2) The assembly process is to perform assembly operations under the condition that there is a workpiece to be assembled on the assembly table and there are small cylindrical workpieces under the assembly manipulator.

In the main program, when the initial state check is over, the unit is confirmed to be ready, and the Run button is pressed to enter the running state, two subroutines of blanking control and assembly control should be called at the same time. The main program calls the assembly program as shown in the ladder diagram in Figure 5-43.

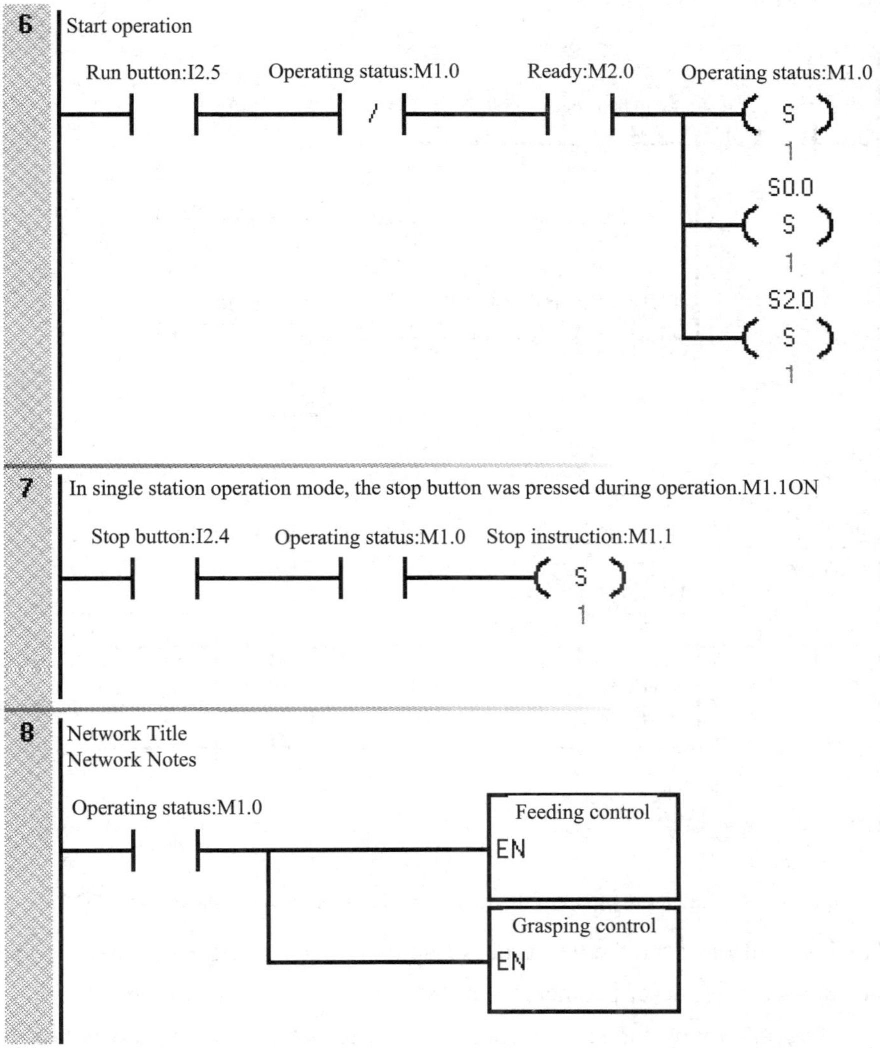

Figure 5-43 Ladder diagram of assembly program calling

The feeding control process includes two interlocking processes, namely the blanking process and the rotation of the swing table and the transfer of the material tray. During the process of the small cylindrical workpieces fall from the silo to the left tray, the swing table is forbidden to rotate; on the contrary, during the rotation of the swing table, it is forbidden to open the silo (the stopper cylinder is retracted) for feeding process. Figure 5-44 shows the ladder diagram of the rotary operation program of the swing cylinder.

The steps to achieve interlocking are as follows:

(1) When the left or right limit solenoid switch of the swing table is activated and the left tray has no cylindrical workpieces, the blanking process will start after timing confirmation.

(2) When the stopper cylinder is extended to the limit position, the silo closes, the left tray has cylindrical workpieces and the right tray is empty, after timing confirmation, the swing table starts to rotate until it reaches the limit position.

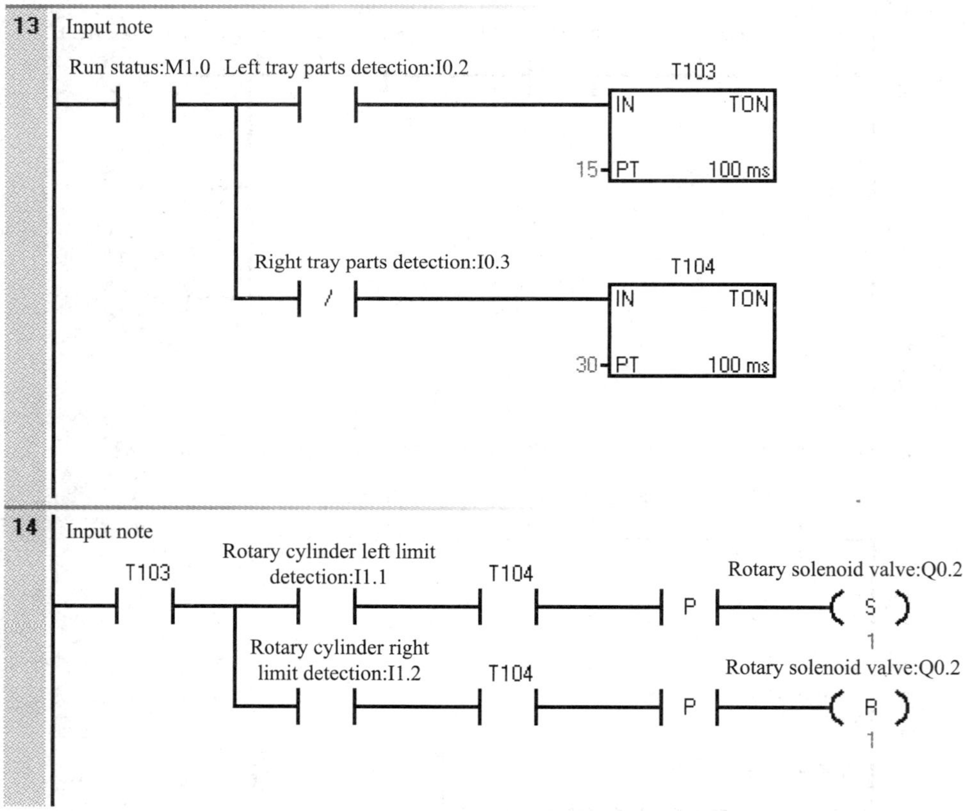

Figure 5-44 Ladder diagram of the rotary operation program of the swing cylinder

The blanking control of the feeding process is all single-step-sequence control. The specific programming steps are not described here. Readers can compile on their own.

There are two situations for stop operation. One is that when the stop button is pressed during operation, the stop command is set; the other is that when the last cylindrical workpiece in the silo falls, the sensor that detects whether the parts (workpieces) are unavailable (I0.1 OFF), if yes, send out a parts unavailable detection alarm.

For the blanking control of the feeding process, under the above two cases the silo should be closed. The ejector cylinder is reset to the limit position (the initial step), and the next blanking is stopped and the blanking is reset to initial step. But for the swing control, once the stop command is issued, the swing should be stopped immediately as shown. For assembly control, the above two cases should also be stopped after one assembly is completed and the assembly manipulator returns to the initial position. Only when the blanking mechanical structure and the assembly manipulator return to the initial position, the running status flag and the stop command can be reset. The operation program for stop running should be programmed in the main program. The ladder diagram is shown in Figure 5-45.

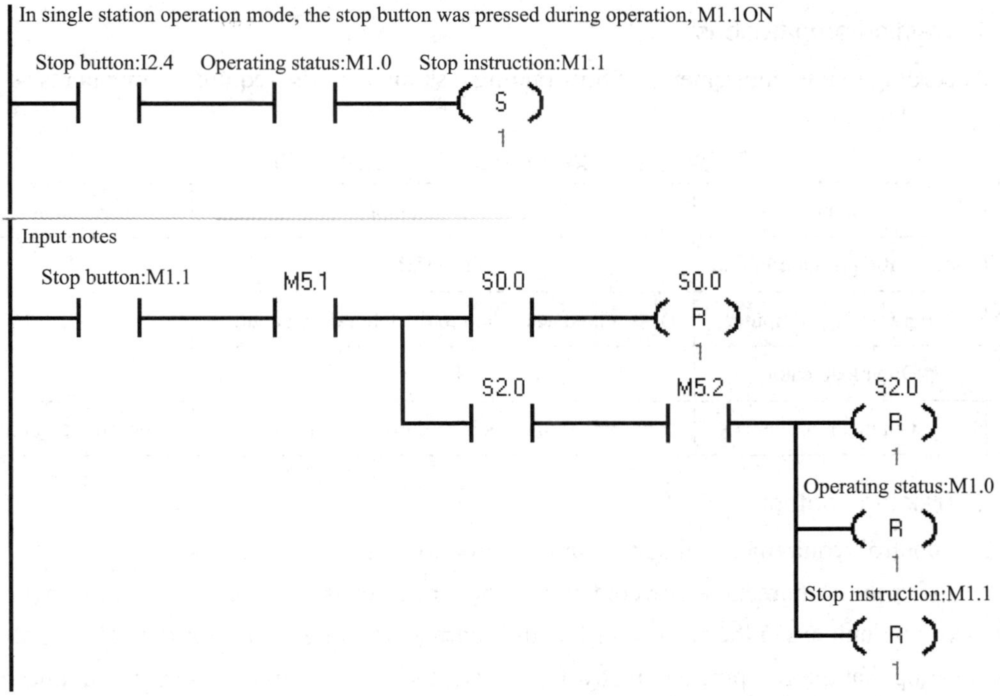

Figure 5-45 Ladder diagram of operation program for stop running

4. Precautions

There are some precautions during compiling the program:

(1) It is necessary to ensure that the blanking process, the rotation of the swing table, and the transfer of the material tray are interlocked.

(2) It is necessary to consider the two cases of stopping when programming. One is stop after finishing one assembly process; the other is stop after the assembly manipulator return to the initial position.

5. Notes and summary

【Extended Practice】

Exercise 1 Programming and debugging of the sorting unit control system

1. Training preparations

According to the requirements of this training task, prepare the required components as shown in Table 5-8.

Table 5-8 Component Preparation List

No.	Name	Specification model	Amount
1	Automatic production line	YL-335B	1
2	Programming computer	Determined according to the laboratory situation	1
3	Download cable	PPI cable	1
4	Connection line	YL-335B special training equipment	Appropriate amount

2. Training content

The control requirements of sorting unit control program are as follows.

(1) After the equipment is powered on and the air source is connected, if the three cylinders of the working unit are in the retracted state, the "normal working" indicator light HL1 is always on, indicating that the equipment is ready. Otherwise, the indicator light flashes at a frequency of 1 Hz.

(2) If the equipment is ready, press the run button to start the equipment, and the "equipment operation" indicator HL2 is always on. When the assembled workpieces are manually put down at the feeder of the conveyor belt, the frequency converter is activated, and the drive motor drives the workpieces to the sorting area at a fixed frequency of 30 Hz.

(3) If the workpiece is a white core metal piece, the workpiece pair reaches the middle of the No.1 chute, the conveyor belt stops, and the workpiece pair is pushed into the No.1 slot; If the workpiece is white core plastic piece, the workpiece pair reaches the middle of the No.2 chute, the conveyor belt stops, and the workpiece pair is pushed into the No.2 slot; If the workpiece is a black core metal/plastic piece, the workpiece pair reaches the middle of the No.3 chute, the conveyor belt stops, and the workpiece pair is pushed into the No.3 slot. After the workpieces are pushed on the chute and into the slot, one work cycle of the work unit ends. Only

after the last the work cycle ends, can workpiece be fed to the conveyor belt again.

(4) If the stop button is pressed during operation, the working unit will stop running after the end of the work cycle.

The work requirements of the sorting unit are as follows:

(1) Planning the I/O distribution and wiring terminal distribution of the PLC.

(2) Carry out system installation wiring and gas circuit connection.

(3) Compile PLC program.

(4) Program debugging and running.

3. Task procedures

1) PLC I/O distribution and wiring

The signal table of sorting unit PLC is shown in Table 5-9, and the schematic diagram of I/O wiring is shown in Figure 5-46.

Table 5-9 I/O signal table of the sorting unit PLC

Input signal				Output signal			
No.	PLC input point	Signal name	Signal source	No.	PLC input point	Signal name	Signal source
1	I0.0	Rotary encoder phase B	Device side	1	Q0.0	Motor start	Frequency converter
2	I0.1	Rotary encoder phase A		2	Q0.1		
3	I0.2	Fiber optic sensor 1 control		3	Q0.2		
4	I0.3	Fiber optic sensor 2 control (Material inflow detection)		4			
5	I0.4	Inlet workpiece inspection (Metal detection)		5	Q0.3		
6	I0.5	Inductive sensor detection (White material detection)		6	Q0.4		
7	I0.6			7	Q0.5		
8	I0.7	Push-out in place detection of rod 1		8	Q0.6		
9	I1.0	Push-out in place detection of rod 2		9	Q0.7	HL1	Button/ indicator module
10	I1.1	Push-out in place detection of rod 3		10	Q1.0	HL2	
11	I1.2	Stop button	Button/ indicator module				
12	I1.3	Run button					
13	I1.4	Emergency stop button					
14	I1.5	Single station/full line					

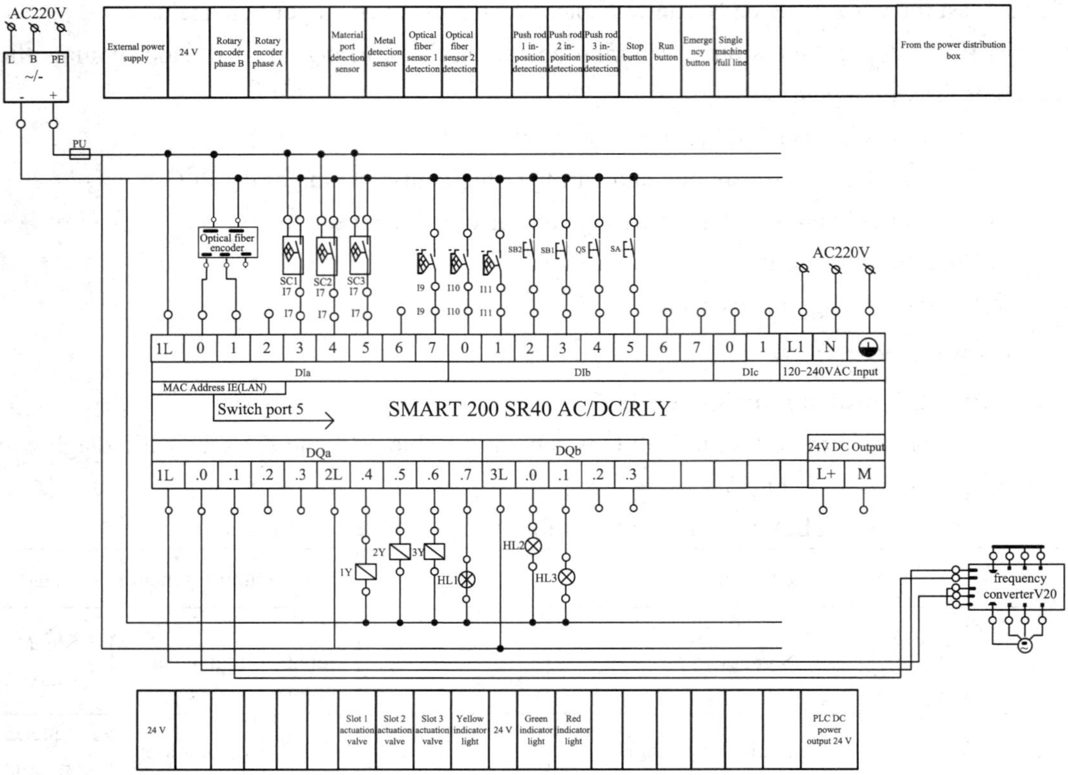

Figure 5-46 I/O wiring schematic diagram of the sorting unit PLC

2) Parameter setting of transfer mechanical structure of sorting unit

The transfer mechanical structure of the sorting unit uses a three-phase asynchronous motor to transfer, and a frequency converter is used for speed regulation. The frequency setting value of the frequency converter is input by the PLC through the D/A conversion mode to input the analog frequency command to realize the continuous adjustment of the motor speed.

3) PLC control programming and debugging

The sorting control subroutine is a step-sequence control program. The programming idea is as follows.

(1) When it is detected that the workpiece to be sorted is discharged to the feed inlet, the current value of HC0 is cleared, and the inverter drive motor is started at a fixed frequency. The ladder diagram program is shown in Figure 5-47.

(2) When the workpiece passes through the optical fiber probe and the inductive sensor mounted on the sensor bracket, the properties of the workpiece are judged according to whether the two sensors are in operation or not, and the flow direction of the program is determined. The comparison between the current value of HC0 and the position value of the sensor can be realized by the contact comparison command. The ladder diagram program to complete the above functions is shown in Figure 5-48.

(3) According to the properties of the workpiece and the requirements of the sorting task, the workpiece is pushed out when it is at the corresponding position of pusher cylinder. After the

material pusher cylinder returns, the step sequence control subroutine returns to the initial step. Readers are requested to complete this part of the program by themselves.

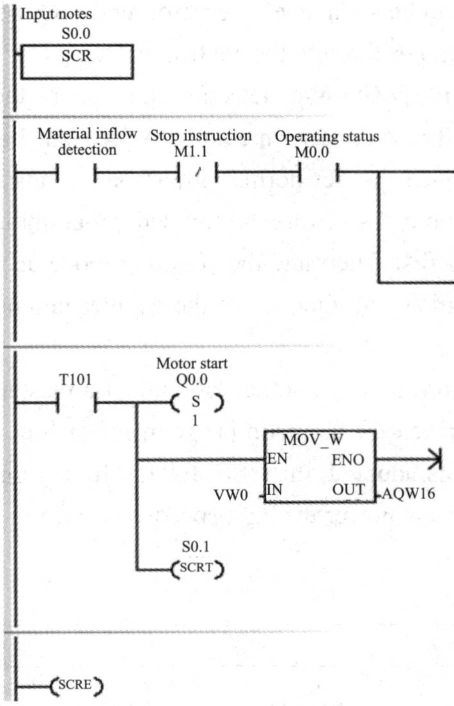

Figure 5-47 Ladder diagram of the initial steps of the sorting control subroutine

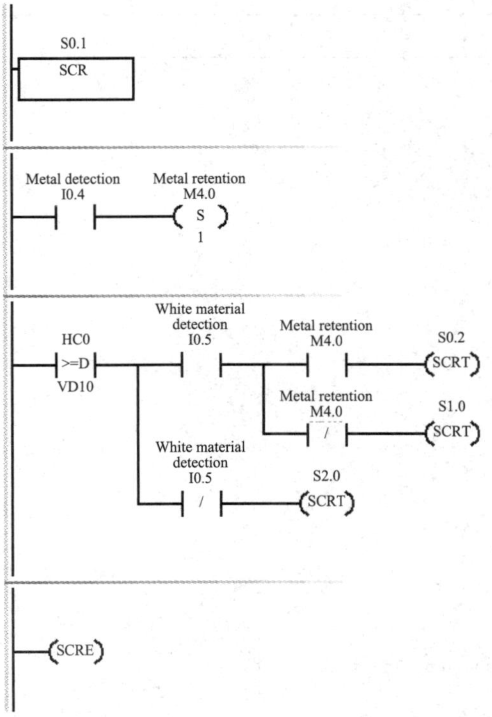

Figure 5-48 Ladder diagram for identifying workpiece attributes at the sensor position

4. Precautions

Precautions during programming are as follows.

(1) The inverter parameters need to be set in analog control mode, so the analog values can control the operation of the conveyor motor through the control of inverter.

(2) The sorting unit needs to use the positioning detection function, so the program needs to use high-speed counter instructions. There are two methods to program high-speed counters. One is to use ladder diagram or statement list for normal programming, and the other is to use STEP7-Micro/WIN SMART programming software for guided programming. Regardless of which method is adopted, you should first determine the counting mode according to the form and requirements of the counting input signal, then select the counter number to determine the input address.

(3) The control program of the sorting unit also needs to adopt a master-slave structure. A sorting control subprogram can be written for the main program for calling. However, because the high-speed counter programming is adopted, the HSC_INIT subprogram must be called in the first power-on scan cycle to define and power the high-speed counter.

5. Notes and summary

____

____

____

____

____

____

Appendix

Appendix 1 Training standards for advanced skills of electricians

1. Overview of Training

According to the *National Vocational Qualification Standard for Electronics*, combined with the employment needs of local people in Cambodia and Chinese-funded enterprises, technical skills training will be carried out by relying on the Lancang-Mekong vocational education training center and experimental training equipment of Luban Workshop in Cambodia.

2. Objectives of Training

Local people in Lancang-Mekong countries, employees of Chinese-funded enterprises, and those who are engaged in or are planning to engage in the installation, debugging, operation, maintenance and repair of mechanical equipment and electrical system lines and devices.

3. Design of Training

1) Overall of design

The advanced electrician training is designed for a total of 120 hours, divided into five learning and training modules: maintenance and troubleshooting of common electrical control circuits, comprehensive training of power electronic circuit technology, comprehensive training of electric drive system, comprehensive training of programmable logic control system, programming and debugging of YL-335B automatic production line, 2–4 tasks are designed. Each task is planned to train 4–12 hours, covering the theoretical knowledge required for the task and practical operation training. At the same time, each training module provides extension training tasks for well-grounded students as further practice after class.

2) Basis of Design

According to the *National Vocational Qualification Standard for Electronics* (occupation code 6-31-03-01) issued by the Ministry of Human Resources and Social Security of the People's Republic of China and the syllabus of Tianjin Sino-German University of Applied Sciences, the training pays full attention to various requirements for the current level of social production development of practitioners.

3) Thoughts on content arrangement

This advanced training course adopts the teaching method of combining multimedia teaching with experimental teaching. According to the characteristics of each module, heuristic teaching runs through the whole course. Teachers are required to pay full attention to classroom teaching in the teaching process, focusing on how to guide students to analyze and solve problems in an electrical system, combine imparting knowledge with developing ability, for students to deepen a real understanding of knowledge through independent learning, for them to improve professional technology ability, professional working ability and professional accomplishment by solving practical problems.

4. Class schedule

Training module	Task	Credit Hours	Total
Module 1 Maintenance and troubleshooting of common electrical control circuits	Task 1 Maintenance and troubleshooting of common faults in the electrical control circuit of CA6140 lathe	4	12
	Task 2 Maintenance and troubleshooting of common faults in the electric control circuit of electric hoist	4	
	Task 3 Maintenance and troubleshooting of common faults in the electrical control circuit of M7120 surface grinder	4	
Module 2 Comprehensive training of power electronic circuit technology	Task 1 Comprehensive training of three-phase bridge-type fully-controlled rectifier circuit	8	24
	Task 2 Comprehensive training of three-phase AC voltage regulating circuit	8	
	Task 3 Comprehensive training of PWM controlled switching regulated power supply	8	
Module 3 Comprehensive training of electric drive system	Task 1 Basic training of fast speed regulation of MM420 inverter	8	24
	Task 2 Basic training of multi-stage speed regulation of MM420 inverter	8	
	Task 3 Training of analog smooth speed regulation of MM420 inverter	8	
Module 4 Comprehensive training of programmable logic control system	Task 1 Programming and debugging of automatic feeding and loading control system	8	24
	Task 2 Programming and debugging of traffic light automatic control system	8	
	Task 3 Programming and debugging of four-floor elevator control system	8	
Module 5 Programming and debugging of YL-335B automatic production line	Task 1 Programming and debugging of feeding unit control system	12	36
	Task 2 Programming and debugging of processing unit control system	12	
	Task 3 Programming and debugging of assembly unit control system	12	
Total class hours	120		

4. Training objectives

1) Overall objective

This training is to design 5 typical training modules and 15 work tasks for the practitioners of senior electrician occupation. The training targets are mainly for circuit maintenance and troubleshooting, circuit debugging, inverter use, and PLC programming. Involved content includes common electrical control circuits, power electronic circuits, MM420 general frequency inverter, S7-200 PLC programming debugging, YL-335B automatic production line debugging, etc. Through the training of this project, trainees can meet the skill requirements of senior electrician. The trainees have the technical theory and operation skills required by senior workers of this type in *Standard for Workers' Technical Grade* and can complete the routine work of this type of work independently using basic skills.

2) Specific objectives

a. Professional competence

① Have a good sense of safety in production and be able to consciously operate according to regulations.

② Have environmental awareness and safety awareness, and be able to consciously keep the workplace clean and tidy.

③ Have the ability to analyze and solve problems.

④ Have a good teamwork spirit, and actively adapt to teamwork requirements.

⑤ Have good professional ethics and social responsibility.

⑥ With self-learning ability, independent working ability and cooperation ability.

b. Methodological competence

① Have the ability to deal with the failure of the electrical control circuit.

② Have the ability to consult manuals and other reference books and equipment nameplates, product manuals, product catalogs and other materials.

③ Have the ability to draw PLC electrical control schematic diagram.

④ Have the ability to select the parameters of commonly used low-voltage electrical appliances according to needs.

⑤ Have the ability to design the control circuit and choose the best control circuit scheme according to the given control requirements.

⑥ Have the ability to use instruments and meters.

⑦ Have the ability to perform correct operations in accordance with operating specifications.

⑧ Have the ability to correctly record and analyze various inspection results.

⑨ Have the ability to use S7-200 programmable controller correctly.

⑩ Have the ability to debug automatic production line equipment.

c. Social competence

① Have good self-cultivation and professional ethics, and gradually develop a persevering learning spirit.

② Have good psychological quality and the ability to overcome difficulties.

③ Care about the current status and trends of scientific and technological development at home and abroad and have a strong sense of mission and responsibility.

④ Have a sense of safety and environmental protection.

5. Training Contents and Requirements

This training module is based on the advanced level of electrician. According to the requirements of different training plans, a series of trainings composed of the merger and reorganization of certain knowledge modules, and the connotation, training focus and hours of each module are also different in different trainings. The senior electrician mainly uses the training equipment of Zhejiang Yalong Educational Technology Equipment Group as the carrier. Each module is equipped with sufficient task content, and a certain amount of expansion exercises are also arranged after the training. The purpose of training is not only to help students consolidate and deepen their understanding of the theoretical knowledge they have learned, but more importantly, to train their practical skills, to establish practical engineering viewpoints and rigorous scientific style. During the training process, the trainees are required to participate in the study and training in accordance with the training plan, and undergo the completion assessment, and the training certificate can be obtained after passing the assessment.

6. Suggestions on Implementation

1) Suggestions on organization of the training

According to the needs of overseas technical training, local government or relevant departments will recruit students, and the Chinese and Cambodian teachers of Lancang-Mekong project team will prepare lessons according to specific training requirements.

2) Suggestions on selection of textbooks

The special training textbook *Lancang-Mekong Vocational Education and Training Center Project Group* are recommended.

3) Suggestions on training methods

Integrated training and learning of theory and practice is carried out based on the systematization of the work process and task introduction method.

4) Suggestions on the development and utilization of training resources

Make full use of the training equipment of the Lancang-Mekong Vocational Education and Training Center to carry out targeted technical skills training, and adopt project-specific training materials. At the same time, accumulate experience in the training process to improve or supplement the design of training modules and tasks.

5) Suggestions on configuration of experimental training

In Lancang-Mekong Vocational Education Training Center, Power Electronics and Electrical Drive Technology Training Room, Electrical Control Technology Training Room, PLC and Sensor Application Technology Training Room are used for advanced electrician training. Relevant equipment is prepared according to the training instruction materials.

7. Assessment and Evaluation

1) Outline of assessment and evaluation

The training assessment should be carried out according to characteristics of the trainees and focus on the basic operating skills. The traditional assessment methods should be reformed, and a scientific, flexible, and usable project evaluation system should be established. The knowledge and technical ability and professional ability are comprehensively evaluated by the process assessment method of observation, operation, record, discussion, group evaluation and skill test.

2) Design of assessment and evaluation

(1) Attendance and in-class questions account for 10% of the total assessment result and the final assessment result accounts for 70%.

(2) The final assessment is based on the combination of theory and practice. Close-book examination is used for theoretical knowledge, accounting for 50% of the final assessment result; task-based training is used for practical operation skills accounting for 50% of the final assessment result.

(3) The total score = 10% of the attendance + 90% of final assessment score.

Appendix 2 Question bank of electronics for advanced level (with answers)

Advanced Electrical Engineering Theory Examination Paper (A)

I. Multiple-choice Questions (the following questions have multiple options, and only one of which is correct. Please choose the correct answer. There are 40 questions in this part. 1 point for each question, and 40 points in total)

1. In the circuit or inside the inverter that is supplied by the power supply to the inverter, when an overcurrent or short-circuit accident occurs, in order to prevent the system from spreading, and at the same time to minimize the damage to the internal components of the inverter, the power supply side must be installed ( ) as the system protect.

   A. contactor B. circuit breaker C. filter D. capacitor

2. The temperature of the elevator deceleration oil tank shall not be higher than ( ) ℃.

   A. 60 B. 50 C. 85 D. 105

3. Double microcomputer control mode elevators are mainly composed of the CPU of the control system, the CPU of ( ) and some relays.

   A. electrical system B. circuit system C. drag system D. communication system

4. The newly put into operation transformer shall be subjected to ( ) impact tests.

   A. 1 B. 3 C. 5 D. 7

5. The "temperature rise" on the motor nameplate refers to the allowable temperature rise of ( ).

   A. stator winding B. stator core C. rotor D. cooling air temperature.

6. During the transformer withstand voltage test, the voltage duration is ( ) minutes.

   A. 1 B. 2 C. 3 D. 5

7. The elevator inspection speed is not allowed to exceed ( )m/s.

   A. 0.25 B. 0.5 C. 0.63 D. 1

8. The carrier frequency has a greater influence on the motor current. When the frequency increases to what level, the motor current waveform is smoother? ( )

   A. above 500 Hz B. above 1.5 kHz C. above 5 kHz D. above 10 kHz

9. When the power system fails, the relay protection of this line is required. The action that "perform those should be performed, and do not perform those that should not be performed" is called the ( ) of relay protection.

   A. selectivity B. sensitivity C. reliability D. rapidity

10. The part of a computer instruction that specifies its action is called ( ).

    A. source address code B. operation code

C. destination address code D. data code

11. Hand-held illuminators and local illuminators used in environments with electric shock hazards should use ( )V safety voltage.

A. 12 B. 6 C. 36 or 24 D. 48

12. The dynamic characteristic index of the sensor is ( ).

A. repeatability B. linearity C. sensitivity D. natural frequency

13. Classified according to the working principle, the solid image sensor belongs to ( ).

A. photoelectric sensor B. capacitive sensor

C. piezoelectric sensor D. magnetoelectric sensor

14. The most obvious feature of vector control system is ( ).

A. dynamic high-speed response B. stepless speed regulation

C. reduce motor loss D. low price

15. Which of the following characteristics is not a characteristic of the intelligent module IPM? ( )

A. It contains over-current, short-circuit and under-voltage protection circuits. Since the detection is carried out internally, it is very convenient and greatly improves the sensitivity and accuracy of the detection

B. Built-in overheat protection to improve the sensitivity of protection

C. Poor heat dissipation

D. Further reduced the size

16. Insulating gloves are easy to break and age. Therefore, their inspection and pressure test should be carried out every ( ) months.

A. 2 B. 6 C. 3 D. 12

17. When calling a function block in the ladder diagram, the box is the function block ( ), and the outside of the box is the corresponding ( ).

A. formal parameters, formal parameters B. actual parameters, actual parameters

C. formal parameters, actual parameters D. actual parameters, formal parameters

18. When the PLC CPU detects an error, if there is no corresponding error handling OB, the CPU will enter ( ) mode.

A. stop B. run C. alarm D. interrupt

19. When the motor winding is disconnected, you should ( ).

A. shut down and check B. observe the situation

C. continue to run

20. Multi-strand soft bare copper wire shall be used for the installation of the grounding wire, and the cross-section of it shall meet the requirements of short-circuit current, but shall not be less than ( ) mm^2.

A. 25 B. 16 C. 50 D. 6

21. The definition of modulation ratio is ( ).

A. in the SPWM inverter, the ratio of the triangle wave voltage frequency to the modulation wave voltage frequency is the modulation ratio

B. in the SPWM inverter, the ratio of the modulating wave voltage frequency to the triangular wave voltage frequency is the modulation ratio

C. in the SPWM inverter, the ratio of the input voltage frequency to the inverter output frequency is the modulation ratio

D. in the SPWM inverter, the ratio of the inverter output frequency to the input voltage frequency is the modulation ratio

22. The elevator motor can not only meet the requirements of frequent starting and braking, but also the requirements (　).

A. large starting current　　B. mechanical characteristics are soft

C. great starting torque　　D. small starting torque

23. The resistance of a conductor is a property of the conductor itself. Which of the following statements is wrong? (　)

A. It is related to the conductor area

B. It is related to the conductor length

C. It has nothing to do with the ambient temperature

D. It is related to the material properties

24. The English abbreviation of sine pulse width modulation is (　).

A. PWM　　B. PAM　　C. SPWM　　D. SPAM

25. The depth of directly buried cables shall not be less than (　)m.

A. 0.5　　B. 0.7　　C. 1　　D. 2

26. The transformation ratio of the transformer is (　).

A. the turns ratio of the high and low voltage windings

B. the ratio of the rated voltages of the high and low voltage sides at no load

C. at rated load, the ratio of the rated voltages on the high and low voltage sides

27. The characteristics of the vacuum circuit breaker are that the high insulation strength of the vacuum medium, the extinguishing capacity is (　), and the service life of the arc energy small circuit breaker is (　).

A. weak; short　　B. weak; long　　C. strong; short　　D. strong; long

28. The "or" logic in digital logic corresponds to (　) of the relay control circuit.

A. series connection　　B. parallel connection

C. bridge connection　　D. series and parallel connection

29. Usually the no-load current I0 of the transformer is (　).

A. AC active current

B. AC reactive current

C. the phasor sum of the active and reactive components of AC current

D. DC current

30. The control signal of the magnetic head of the single electronically controlled solenoid directional valve is usually required (　).

A. pulse signal　　B. continuous signal　　C. delay signal　　D. nothing

31. When measuring equipment insulation with a megger, the speed of the handle should be close to (   )r/min.

A. 100   B. 120   C. 150   D. 180

32. When arranging fire-fighting equipment, the distance between the control panel and the wall should not be (   ) 3 m on the side where the duty personnel often work.

A. less than   B. greater than

C. equal to   D. greater than or equal

33. The RST pin of the single-chip microcomputer needs to input a high level signal that maintains more than (   ) machine cycles to be reset.

A. 2   B. 4   C. 6   D. 8

34. There is a three-phase asynchronous motor with Y-connection in operation. If its line current is 10A, when the two phases are clamped with a clamp-type ammeter, the reading should be (   )A.

A. 10   B. 20   C. 17.3   D. 5.77

35. In a sinusoidal AC circuit, keep the voltage across the capacitor unchanged, reduce the frequency of the power supply, and the current flowing through the capacitor will be (   ).

A. increase   B. decrease   C. no change   D. uncertain

36. Explosion-proof vacuum feed switch, vacuum circuit breaker and isolating switch are installed in (   ), and the control unit is installed on the door cover.

A. main cavity   B. door cover

C. junction box   D. vacuum circuit breaker

37. The high-voltage vacuum contactor is suitable for (   ) three-phase AC 50 Hz power system.

A. 1−2 kV   B. 3−10 kV   C. 11−12 kV   D. 13−15 kV

38. In the hydraulic system shown on the right, the adjustment pressures of the two overflow valves are known to be: $P_y1=50\times10^5$ Pa and $P_y2=20\times10^5$ Pa. When the piston moves to the left and right, the maximum pressure that the hydraulic pump can reach: to the left is (   ), and to the right is (   ).

A. 50 bar   B. 20 bar

C. 30 bar   D. 70 bar

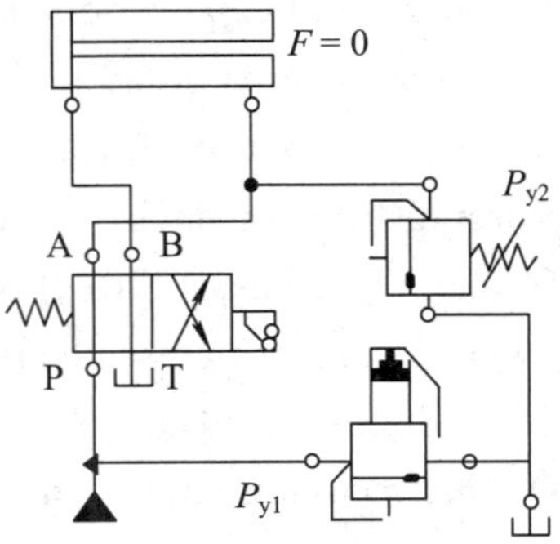

39. When the series circuit of R, L, C resonates, the resonant frequency is (   ).

A. Only related to R and L

B. Only related to R and C

C. Only related to R

D. Only related to L and C, not related to R

40. When the triode constitutes the amplifier, there can be (   ) kinds of connection methods according to the difference of the common terminal.

A. 1   B. 2   C. 3   D. 4

Ⅱ. Multiple-Answers Questions (the following questions have multiple options, of which two or more options are correct. Please choose the correct answer. There are 10 questions. 2 points for each question, and 20 points in total)

1. The conditions for scrapping the wire rope including ( ).
   A. the diameter is less than 90% of the original diameter
   B. there are 2–3 broken wires in one lay length
   C. use for more than one year
   D. monofilament wear exceeds 40% of original diameter
2. The safe operation device to protect passengers including ( ).
   A. car top inspection switch B. safety touch panel
   C. hall door and interlock switch D. safety gear switch
3. Frequency conversion speed regulation system can generally be divided into ( ) category.
   A. AC-AC frequency conversion B. direct-AC-direct frequency conversion
   C. direct-AC frequency conversion D. AC-DC-AC frequency conversion
4. When the maintenance personnel go down the pit of the shaft, the driver's operation requirements are ( ).
   A. turn off the power switch and maintenance personnel enter the pit
   B. the elevator runs to the end of the station
   C. the elevator leaves about 2 m from the bottom station
   D. turn on the maintenance switch, maintenance personnel enter the pit
5. The 8051 MCU ( ) port does not have an internal pull-up resistor.
   A. P0 B. P1 C. P2 D. P3
6. The influence of AC negative feedback on the amplifying circuit has ( ).
   A. stable magnification B. increase the input resistance
   C. improve distortion D. stable static operating point
7. When overhauling high-voltage motors and starting devices, the following safety measures should be taken: ( ).
   A. disconnect the power circuit breaker (switch) and isolating switch (knife switch); install a grounding wire or install an insulating partition between the isolating switch (knife switch) after the experience is clear, and the trolley switch should be pulled out from the complete power distribution device and close the door and lock it
   B. hang the signs of "No closing, someone working!" on the handles of the circuit breaker (switch) and isolating switch (knife switch)
   C. the disconnected cable head must be three-phase short-circuited to ground
   D. take measures to prevent the driven machinery (such as water pumps, air compressors, induced draft fans, etc.) from causing the motor to rotate, and hang a sign "No closing, someone working!" on the valve.
8. Commonly used control methods of bidirectional thyristors are ( ).

A. sharp pulse trigger
B. phase shift trigger
C. zero-cross trigger
D. frequency conversion trigger

9. When the transmission line of the automatic fire alarm system adopts insulated wires, it should be wired with ( ) or enclosed trunking.

A. flame-retardant rigid plastic tube
B. flame-retardant semi-rigid plastic tube
C. rubber tube
D. metal tube

10. The main technical parameters of vacuum circuit breakers include impulse insulation voltage level, power frequency withstand voltage, mechanical life and ( ), etc.

A. rated voltage
B. rated current
C. short-circuit breaking voltage
D. short-circuit closing voltage

Ⅲ. True or False (there are 30 questions, 1 point for each question, and 30 points in total)

1. The iron loss of the transformer refers to the loss in the winding of the transformer. ( )
2. When the amplifier is working, there are both DC and AC components in the circuit. ( )
3. The PLC power module only has the function of short-circuit protection, excluding the power-down protection circuit and the backup battery power supply. ( )
4. The main factor affecting the viscosity of hydraulic oil is temperature. ( )
5. The capacitive humidity sensor can only measure relative humidity. ( )
6. An instrument factory needs to purchase a pressure gauge. If the relative error of the full scale of the pressure gauge is less than 0.7%, it should buy a pressure gauge of class 1.0. ( )
7. When the ultrasonic wave enters the steel at an oblique angle of 45° from water, the refraction angle is equal to the incident angle. ( )
8. For the use of pneumatic locks in pneumatic systems, a three-position reversing valve with an H-type neutral function should be selected. ( )
9. The background data block can only be given to a single function block, and the same function block can establish an assignment relationship with multiple background data blocks. ( )
10. The selection of PLC digital signal template should pay attention to its voltage level, template density and threshold level. ( )
11. Running under the condition of low frequency *U*/*f*, there is no potential for energy saving. ( )
12. The transformer after overhaul needs to be subjected to 3 shock tests. ( )
13. When using a clamp meter, the two surfaces of the jaws should be in good contact with no impurities. ( )
14. When measuring the insulation with a megger, the E end is connected to the wire, and the L end is grounded. ( )
15. The metal shell grounding of electrical equipment belongs to working grounding. ( )
16. It is strictly forbidden for staff to move or remove fences, grounding wires and nameplates during work. ( )

17. The cooling methods of the transformer include oil-immersed self-cooling, oil-immersed air-cooling, strong oil-air-cooling and strong oil-water cooling. ( )
18. The speed control of the two-speed elevator is open loop control. ( )
19. Check the insulation resistance between the stator and rotor winding phases of the low-voltage motor and the winding to the ground. When measuring with 500 V insulation resistance, the value should not be lower than 0.5 MΩ, otherwise, it should be dried. ( )
20. MCS-51 MCU is a high-end 16-bit MCU. ( )
21. When measuring the capacitance of a power capacitor, if it is found that the actual measured value this time is significantly larger than the last measured value or the value on the nameplate, it may be that there is a breakdown component inside the capacitor. ( )
22. Zinc oxide arresters are different from ordinary valve arresters in that they have no internal gaps and have excellent volt-ampere characteristics. ( )
23. Taking oil samples on oil-filled electrical equipment should be done when the weather is dry. ( )
24. The cement reactor is a hollow coil wound on the cement pillar, and its inductance is a fixed value. ( )
25. The higher the order of the harmonic, the greater the reactance value displayed by the capacitor to the harmonic. ( )
26. The gap between the brake core and the shaft sleeve is very small. When there is dust and rust on the surface, the brake may not operate when the power is cut off, causing serious elevator accidents. ( )
27. When conducting the cable withstand voltage test, an AC high voltage needs to be applied between the cable core and the ground. ( )
28. The new equipment can be put into operation if there is a factory test report. ( )
29. Allow AC and DC circuits to share one cable. ( )
30. To shorten the traction wire rope, it is necessary to re-make the wire rope head and cast Babbitt alloy. ( )

Ⅳ. Short answer questions (there are 2 small questions, each with 5 points, 10 points total)

1. What are the speed regulation methods for DC motors? And point out the speed regulation methods commonly used in practice.
2. What are the common electrical faults of the three-phase cage asynchronous motor itself?

Answers to Advanced Electrical Engineering Theory Examination Paper (A)

I. Multiple-choice Questions

Scoring criteria: points are awarded for correct answers, points are not awarded for incorrect or unanswered questions, and points are not deducted; each question is 1 point, 40 points in total.

1	2	3	4	5	6	7	8	9	10
B	C	C	C	A	A	C	D	B	B
11	12	13	14	15	16	17	18	19	20
C	D	A	A	C	B	C	A	A	A
21	22	23	24	25	26	27	28	29	30
A	C	C	C	B	B	D	B	C	B
31	32	33	34	35	36	37	38	39	40
B	A	A	A	B	A	B	BA	D	C

Ⅱ. Multiple Answers Questions

Scoring criteria: points are awarded for correct answers to each question, and no points are given or deducted for wrong selection, multiple selections, or fewer selections; 2 points for each question, 20 points in total.

1	2	3	4	5	6	7	8	9	10
ABD	BCD	AD	ABCD	BCD	ABC	ABCD	ABC	AD	AB

Ⅲ. True or False

Scoring criteria: points are awarded for correct answers, points are not awarded for incorrect or unanswered questions, and points are not deducted; each question is 1 point, a total of 30 points.

1	2	3	4	5	6	7	8	9	10
×	√	×	×	√	×	×	×	√	√
11	12	13	14	15	16	17	18	19	20
×	√	√	×	×	√	√	√	√	×
21	22	23	24	25	26	27	28	29	30
√	√	√	√	×	×	×	√	×	×

Ⅳ. Short answer questions

1. Methods of speeding regulation

① pressure and speed regulation.

② Magnetic and speed adjustment.

③ Adjust the resistance to adjust the speed.

④ In practice, the method of voltage regulation and magnetic regulation are often used for smooth and stable speed regulation.

Scoring criteria: ①②③④ account for 25% each.

2. The common electrical faults of the three-phase squirrel cage asynchronous motor itself are: ① the winding is grounded; ② the winding is open; ③ the winding is short-circuited; ④ the winding is wrongly connected; ⑤ the reverse is inserted; ⑥ the rotor is broken or the end ring is broken.

Scoring criteria: ①②③④⑤ accounted for 15% each; ⑥ accounts for 25%.

Advanced Electrical Engineering Theory Examination Paper (B)

I. Multiple-choice Questions (the following questions have multiple options, and only one of which is correct. Please choose the correct answer. There are 40 questions in this part. 1 point for each question, and 40 points in total)

1. Electrical workers should take an examination once a year for the *Electrical Industry Safety Work Regulations*. For those who have interrupted electrical work continuously (　) or more, they must review this regulation again and pass the exam before they can resume work.

A. 1 year　B. 3 months　C. 6 months　D. 2 years.

2. Classified according to the working principle, the solid image sensor belongs to (　).

A. photoelectric sensor　B. capacitive sensor

C. piezoelectric sensor　D. magnetoelectric sensor

3. The fundamental reason why the inverter interferes with other equipment is because (　).

A. there are high-order harmonic components in the input current

B. there are high-order harmonic components in the output current

C. there are high-order harmonic components in the input and output

D. no high-order harmonic components in the input and output

4. The number of electricity inspectors (　) when electricity inspectors conduct on-site inspections.

A. at least one person　B. not less than two people

C. not less than three people　D. not less than four people

5. The frequency conversion speed regulation of the motor down from the basic frequency belongs to (　) speed regulation.

A. constant torque　B. constant voltage　C. constant power　D. constant slip

6. When testing on lines with capacitors and cables, you should first fully (　) and then test.

A. clean up　B. charge　C. discharge　D. none of above

7. Power transformer oil plays a role of (　).

A. coil lubrication　B. insulation and anti-rust

C. insulation and heat dissipation　D. none of above

8. The characteristics of insulated gate bipolar transistors including (　).

A. the breakdown voltage and the maximum saturation current of the collector are small

B. switching frequency can reach above 5 kHz

C. switching frequency up to 20 kHz

D. the breakdown voltage can reach 3000 V, and the maximum saturation current of the collector has exceeded 1000 A

9. The basic idea of von Neumann computer mainly refers to (　).

A. binary coding and storage program control　B. binary numbers and storage

C. CPU and memory　D. memory and input and output devices

10. When the ideal current source is open, the inside of the constant current source ( ).

A. there is current, and there is power loss

B. there is no current, and there is no power loss

C. there is current, and there is no power loss

D. there is no current, and there is power loss

11. In steel pipe wiring, the wires in the same AC circuit must pass through ( ).

A. the same steel pipe B. different steel pipes C. any steel pipe

12. ( ) elevator is commonly known as VVVF elevator.

A. AC speed regulation

B. AC frequency conversion and voltage transformation

C. Hydraulic

D. Direct current

13. The rated voltage of the megohmmeter used to measure the insulation resistance of low-voltage power cables is ( ) V.

A. 2500 B. 500 C. 1000 D. 250

14. Based on what control system was PLC developed ( ).

A. Relay control system B. Single chip microcomputer

C. Industrial computer D. Robot

15. What is the RS232 communication distance in the PLC design specification? ( ).

A. 1300 m B.200 m C.30 m D.15 m

16. After the PLC external contact is broken, change it to another good point and then which menu in the software should be used for operation ( ).

A. Find B. Replace C. Find instructions

17. What kind of protection does the thermal relay do for the motor in the circuit? ( ).

A. Short circuit B. Overload C. Overcurrent D. Overvoltage

18. In which way the touch screen communicates with the PLC ( ).

A. communication B. I/O signal control

C. relay connection D. electrical connection

19. The touch screen is used to realize the function of which devices to replace ( ).

A. traditional relay control system

B. PLC control system

C. industrial computer system

D. traditional switch button type operation panel

20. The scan generator in the oscilloscope is actually an ( ) oscillator.

A. sine wave B. multi-harmonic

C. capacitor three-point type D. inductance three-point type

21. If the output voltage amplitude of the PWM pulse width modulation frequency conversion circuit is increased, the control method that can be used is ( ).

A. increase the amplitude of the triangle wave

B. increase the frequency of the triangle wave

C. increase the frequency of the sine modulation wave

D. increase the amplitude of the sine modulation wave

22. The operating voltage of the cable must not exceed (　　) of the rated voltage.

A. 5%　　B. 10%　　C. 15%　　D. 20%

23. Which of the following statements about the Hall sensor is correct? (　　)

A. Hall sensors have the ability to sense magnetic fields under static conditions.

B. When the control current is constant, the Hall potential is proportional to the magnetic induction intensity.

C. When the control current and the magnetic induction intensity both change, the output of the sensor is proportional to the product of the two.

D. When the magnetic induction intensity is constant, the Hall voltage is inversely proportional to the control current.

24. When measuring the DC resistance of a 1000 kVA three-phase power transformer with a DC bridge, the difference between the lines must not exceed (　　) of average value.

A. 5%　　B. 4%　　C. 3%　　D. 2%

25. For a running Y-connected three-phase asynchronous motor, if its line current is 10 A, when the two phases are clamped with a clamp-type ammeter, the reading should be (　　)A.

A. 10　　B. 20　　C. 17　　D. 30

26. During the transformer withstand voltage test, the voltage duration is (　) minutes.

A. 1　　B. 2　　C. 3　　D. 5

27. After the three-phase asynchronous motor is started, it buzzes, but the reason for not turning is (　　).

A. short circuit　　B. voltage too low　　C. phase loss　　D. overload

28. The rated voltage of the megohmmeter used to measure the insulation resistance of 6 kV power cables is (　　) V.

A. 1000　　B. 2500　　C. 500　　D. 250

29. The coupling capable of compensating for poor alignment or deflection of the two shafts is (　) coupling.

A. flange　　B. slide block　　C. sleeve　　D. gear

30. A complete computer hardware system includes (　　).

A. computer and its peripheral equipment　　B. host, keyboard and display

C. digital circuit　　D. integrated circuit

31. The following physical quantity to be measured are suitable for measurement using infrared sensors is (　　).

A. pressure　　B. torque　　C. temperature　　D. thickness

32. The application program of the single-chip microcomputer is generally placed in (　　).

A. RAM　　B. ROM　　C. register　　D. CPU

33. After the inverter is connected to the reactor at the power input side, the waveform of (　　)

can be greatly improved.

A. output current B. input current

C. output and input current D. intermediate DC current

34. If you want to realize the logical "or" function of the signal in the pneumatic system, ( ) can be used.

A. two signals in series B. double pressure valve

C. shuttle valve D. reversing valve

35. The structure of the feed switch is composed of an explosion-proof enclosure, an isolating switch control unit and a ( ).

A. main chamber B. door cover

C. junction box D. vacuum circuit breaker

36. Generally, the number of parallel detectors should be less than ( ), and the relevant requirements are shown in the design drawings.

A. 3 B. 5 C. 6 D. 2

37. Which statement is correct about the pneumatic control system shown on the right? ( ).

A. The piston speed of the cylinder can be adjusted with valve 1.3.

B. When operating 1.2 or 1.1, the cylinder extends.

C. Operate 1.2 and 1.1, and the cylinder will extend.

D. Inching 1.2 and 1.1, and the cylinder will remain in the extended position.

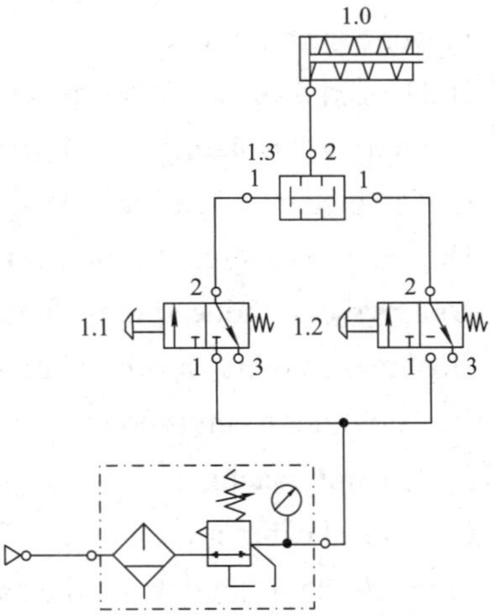

38. The control signal of the solenoid head of the single electric control solenoid directional valve is usually required ( ).

A. pulse signal

B. continuous signal

C. delay signal

D. nothing

39. What signal does the pressure switch convert into what signal? ( ).

A. Electric signal is transformed into pressure signal

B. Gas signal is transformed into electric signal

C. The long signal of qi is converted into a short signal

D. The short signal of electricity is converted into a long signal of qi

40. If you use a measuring instrument to measure and display oil pressure at different positions of the machine, which one of the following instruments will you choose? ( ).

A. Accuracy gauge B. Pressure gauge

C. Pyrometer D. Barometer, barometer

Ⅱ. Multiple-Answers Questions ((the following questions have multiple options, of which two or more options are correct. Please choose the correct answer. There are 10 questions. 2 points for each question, and 20 points in total)

1. The conditions for scrapping the wire rope including: (   ).

   A. the diameter is less than 90% of the original diameter.

   B. there are 2–3 broken wires in one twist pitch

   C. use for more than one year

   D. monofilament wear exceeds 40% of the original diameter

2. Which of the following are proximity switches? (   )

   A. Self-induction   B. Differential transformer type

   C. Eddy current type   D. Capacitive

3. When overhauling high-voltage motors and starting devices, which of the following safety measures should be taken? (   ).

   A. Disconnect the power circuit breaker (switch) and isolating switch (knife switch), install a grounding wire or install an insulating partition between the isolating switch (knife switch) after the experience is clear, and the trolley switch should be pulled out from the complete power distribution device and Close and lock

   B. Hang the signs of "No closing, someone is working!" on the handles of the circuit breaker (switch) and isolating switch (knife switch)

   C. The disconnected cable head must be three-phase short-circuit grounded

   D. Take measures to prevent the motors driven by them (such as water pumps, air compressors, induced draft fans, etc.) from causing the motor to rotate, and hang a sign "No closing, someone working!" on the valve

4. The computer room is composed of (   ).

   A. Traction machine   B. Buffer

   C. Control cabinet   D. Safety gear

5. Which of the following are the basic parameters of the sensor? (   ).

   A. Range   B. Sensitivity   C. Accuracy   D. Voltage

6. The following interrupt sources belonging to 8051 are (   ).

   A. external input interrupt source (P3.2)

   B. external input interrupt source (P3.3)

   C. T0 overflow interrupt source

   D. serial port transmit and/receive interrupt source

7. The main technical parameters of vacuum circuit breakers include impulse insulation voltage level, power frequency withstand voltage, mechanical life and (   ), etc.

   A. rated voltage   B. rated current

   C. short-circuit breaking voltage   D. short-circuit closing voltage

8. The flameproof enclosure of the feed switch is composed of three parts: (   ).

   A. main chamber   B. door cover

C. junction box D. vacuum circuit breaker

9. The automatic fire alarm controller is usually divided into ( ) according to its purpose.

A. regional alarm controller B. centralized alarm controller

C. general alarm controller D. city alarm controller

10. The wiring mode of the automatic fire alarm system is ( ).

A. serial connection method B. trunk connection method

C. star connection method D. delta connection method

Ⅲ. True or False (there are 30 questions, 1 point for each question, and 30 points in total.)

1. One of the conditions for parallel operation of transformers, that is, the primary voltage and secondary voltage of each transformer should be equal, but the allowable error is within ±5%. ( )

2. The IA-32 processor has absorbed the expertise of RISC technology. RISC refers to a complex instruction set computer. ( )

3. The PLC power module only has the function of short-circuit protection, excluding the power-down protection circuit and the backup battery power supply. ( )

4. Inversion is to convert alternating current into direct current. ( )

5. Whether the electromagnetic compatibility of the sensor meets the standard refers to the ability of the sensor to work normally in the specified electromagnetic interference environment. ( )

6. For some equipment with high-frequency interference signals, the equipment grounding also has the function of introducing high-frequency interference signals into the ground. ( )

7. When an accident occurs in the system, the transformer allows overload operation. ( )

8. When using a multimeter to measure resistance, the power must be cut off, but it is not necessary to use a shaker to measure resistance. ( )

9. The advantage of the bridge rectifier circuit is that the output voltage pulsation is small. The back pressure of the rectifier tube is low, and the utilization rate of the transformer is high. ( )

10. The first type of work ticket should be used for power cable outage work, and the second type of work ticket should be used for work that does not require a power outage. ( )

11. In the three-phase half-wave controllable rectifier circuit, the reverse peak voltage that the thyristor may bear is $\sqrt{2}U_2$. ( )

12. The thyristor is a controllable rectifier device composed of four layers of semiconductors with three PN junctions. ( )

13. In the event of a personal or equipment accident, immediately stop the elevator and cut off the power supply, assist in rescuing the injured, report to the relevant department, and protect the scene. ( )

14. Running under the condition of low frequency U/f, there is no potential for energy saving. ( )

15. High-voltage electroscopes can be used for low-voltage electrical inspections, while low-voltage electroscopes cannot be used for high-voltage electrical inspections. ( )

16. The elevator that runs automatically is maintained by a maintenance unit and does not require management personnel. ( )

17. The function of the gearbox is to reduce the speed and increase the torque. ( )

18. One of the functions of the brake is to stop the traction machine and stop the elevator when the elevator encounters an emergency. ( )

19. When the speed is adjusted below the rated frequency, the magnetic flux is kept constant and the voltage-frequency ratio is constant. ( )

20. Special equipment users should use special equipment that meets the requirements of safety technical specifications. ( )

21. Special equipment that has not been inspected regularly or is unqualified may be supervised for use. ( )

22. The actual amplifier circuit often uses a voltage divider bias circuit because of its large input impedance. ( )

23. The power field effect tube is a current-controlled device. ( )

24. Anyone who engages in the manufacture, installation and transformation of special equipment without permission shall be fined 50 000 to 200 000. ( )

25. The copper loss of the transformer refers to the loss in the transformer core. ( )

26. The relationship between each input and output in the logic control system is a pure logic relationship. ( )

27. The higher the order of the harmonics, the greater the reactance value displayed by the capacitor to the harmonics. ( )

28. Too high voltage can produce dangerous temperature, and too low voltage will not produce dangerous temperature. ( )

29. PLC allows dual-coil output during programming (different outputs with the same number). ( )

30. Reasonable wiring can eliminate the intensity of interference signals. ( )

Ⅳ. Short answer questions (2 small questions, 5 points for each, and 10 points in total)

1. What are the advantages and disadvantages of the AC-DC-AC inverter speed control system? What is the application prospect?

2. What problems should be paid attention to when choosing a time relay in the control circuit of a machine tool?

Answers to Advanced Electrical Engineering Theory Examination Paper (B)

Ⅰ. Multiple-Choice Questions

Scoring criteria: points are awarded for correct answers; points are not awarded for incorrect or unanswered questions; and points are not deducted. 1 point for each question, 40 points in total.

1	2	3	4	5	6	7	8	9	10
B	A	C	B	A	C	C	A	A	C
11	12	13	14	15	16	17	18	19	20
A	B	B	A	D	B	B	A	D	A
21	22	23	24	25	26	27	28	29	30
D	C	B	D	A	A	C	B	B	A
31	32	33	34	35	36	37	38	39	40
C	B	B	C	D	B	C	B	B	B

Ⅱ. Multiple-Answers Questions

Scoring criteria: points are awarded for correct answers to each question, and no points are given or deducted for wrong selection, multiple selections, or fewer selections. 2 points for each question, 20 points in total.

1	2	3	4	5	6	7	8	9	10
ABCD	ABCD	ABCD	AC	ABC	ABCD	AB	ABC	ABC	ABC

Ⅲ. True or False

Scoring criteria: points are awarded for correct answers; points are not awarded for incorrect or unanswered questions, and points are not deducted. 1 point for each question, 30 points in total.

1	2	3	4	5	6	7	8	9	10
√	×	×	×	√	√	√	×	√	√
11	12	13	14	15	16	17	18	19	20
×	√	√	×	×	×	√	√	√	×
21	22	23	24	25	26	27	28	29	30
×	×	×	√	×	√	×	×	×	×
31	32	33	34	35	36	37	38	39	40
√	×	×	×	√	×	×	√	×	√

Ⅳ. Short answer questions

1. ① AC-DC-AC variable frequency speed regulation can realize smooth stepless speed regulation, wide speed regulation range, high efficiency, and can give full play to the advantages of three-phase squirrel cage asynchronous motors.

② The disadvantage is that the frequency conversion system is complicated and the cost is high.

③ As the thyristor frequency conversion technology is becoming more and more perfect, its application prospects are promising, and there is a trend to gradually replace the DC motor speed control system.

Scoring criteria: ①② account for 30% each; ③ accounts for 40% each.

2. Select the time relay, the issues should be paid attention:

① mainly consider the delay mode of the delay contact caused by the control circuit; ② the delay range; ③the type of instantaneous contact; ④ the number of contacts; ⑤ the voltage level of the solenoid coil.

Scoring criteria: ①②③④⑤ account for 20%.

Reference

[1] China Employment Training and Guidance Center. National Vocational Qualification Training Course: Maintenance Electrician (Advanced) [M]. 2nd ed. Beijing: China Human Resources & Social Security Publishing Group, 2013.

[2] LI Z, ZHANG S. Automatic production line installation and commissioning [M]. Beijing: China Machine Press，2019.

[3] LIAO C. PLC programming and application [M].4th ed. Beijing: China Machine Press，2014.

[4] TANG X, JIANG Y. Installation and debugging of modern electric control system [M]. Beijing: China Railway Publishing House, 2017.

[5] TANG H, YANG D. Motor drag and frequency conversion speed regulation [M]. Shanghai: East China Normal University Press, 2014.

[6] ZHANG L, YU L, ZHANG Z. Machine tool electrical circuit installation and maintenance [M]. Beijing: China Machine Press, 2019.